ULTRAMICROELECTRODES

Martin Fleischmann
The University, Southampton

Stanley Pons
University of Utah

Debra R. Rolison
Naval Research Laboratories

Parbury P. Schmidt
Oakland University

♦♦♦
Datatech Systems, Inc.
Science Publishers

Orders and correspondence should be directed to

The Editor
DATATECH SYSTEMS, INC.
Scientific Publishing Division
P.O. Box 435
Morganton, NC 28655-0435

Production and Copy Editor
        Sheila B. Pons
Distribution and Sales
        C. Gary Triggs
Design Consulting
        T. Triggs

Copyright ©1987 by Datatech Systems, Inc.

All rights reserved.

No part of this book may be reproduced by any means, nor transmitted, nor translated into a machine language without the written permission of the publisher.

Datatech Serial Publication 0.003.0004.June.30.1987.004.0.0

The Cloth Edition: ISBN 0-9618927-0-6

Lithographed in the USA by Publishers Press

The authors dedicate this book to the Office of Naval Research which has, for so many years, stimulated new fields of research in electrochemistry.

Contents

Preface x
List of Figures xii
List of Tables xvi
List of Participants xvii

1
Introduction and Overview   1
New studies with ultramicroelectrodes   1
Studies in the liquid solution phase   2
    Studies of fast electrode reactions and of fast
        homogeneous reactions   2
    The electrochemical study of reactions in
        nonconducting liquid solutions; catalysis . .   4
    Dispersions of ultramicroelectrodes: the study of
        nonelectrochemical catalysis   4
    Dispersions of ultramicroelectrodes: the study and
        implementation of efficient electrochemical
        syntheses   6
Studies in the solid solution phase   7
    Fundamental electrochemical studies in the solid
        solution phase   7
    Membrane coated ultramicroelectrodes   8
The gas phase   9
    Gas phase electrochemistry and the study of reaction
        kinetics   9
    Catalysis at ultramicroelectrodes   10
    Steam systems   10
Theoretical considerations   11
    Investigation of electrochemical kinetics   11
    Theoretical treatment of the electrical double layer:
        investigations with the use of
        ultramicroelectrodes   12
Miscellaneous applications and problems for investigation
    New constructions and novel systems   13
    Sheets of imbedded ultramicroelectrodes   13
    The fabrication and coating of very small dimensioned
        disk and ring ultramicroelectrodes   14
    The design of experiments with the dropping and
        sessile drop mercury ultramicroelectrode . . .   15

2
Selected Theoretical Topics   17
   Electrochemical behavior of disks and rings 17
      General 18
      The disk 20
      The ring 24
   Polarization curves 28
      Disk electrodes 28
      Ring electrodes 30
   Approximate treatments of experimental data 37
   Extension of the approach to the discussion of the non-
      steady state behavior:
      Chronopotentiometry 40
      Chronoamperometry 47
   Other applications 48
      Reactions in solution coupled to electrode processes
         The c.e. reaction 48
      Linear sweep amperometry 52
      The A.C. impedance 52
   Conclusion 58

3
Less is More: Fabrication of Ultramicroelectrodes   65
   Introduction 65
   The ultramicroelectrode jungle 66
   Micro to macro transitions 72
   The conductor-glass interface 72
   The conductor-epoxy interface 74
   Cleaning/pretreating ultramicrosurfaces 79
   What you see 81
   How much less??? 84
   "Wireless" electrochemistry 85
   Electrochemical microscopy 86
   Appendix 91
      Mercury microspheres 91
      Microcylinder - microdisk 91
      Microdisk - microcylinder arrays 94
      Microring 96
      Band or line 98
      Line arrays 100
      Irregular arrays of irregularly shaped
         microelectrodes 105

4
Selected Reviews   107

James L. Anderson	107
Allen. J. Bard	114
Alan M. Bond	115
Martin Fleischmann	120
Michael Gratzel	123
Jonathon O. Howell	125
Richard McCreery	126
Marcin Majda	127
Barry Miller	129
Royce Murray	131
Janet Osteryoung	136
Vernon Parker	139
Stanley Pons	140
Attilla Szabo	143
Dennis Tallman	145
Wolfgang Thormann	148
Mark Wightman	152

5
Contributions   157

H.D. Abruña	158
Christian Amatore	169
Some properties of ultramicroelectrodes and their implications	169
Characteristic dimension	169
High current densities	170
Nonuniform current densities	171
Ohmic drop	171
Arrays	172
Small geometric size	173
The behavior of ultramicroelectrodes chemically implanted in living tissue	175
Electrochemistry at cylindrical and band electrodes Cyclic voltammetry	177
Conclusion	182
James Anderson	184
New opportunities made possible by microelectrodes	184
Electrochemistry in "unusual" places	184
Microsensors	185
Investigation of very rapid events	185
Monitoring processes at the molecular level	185
New problems to solve	185
Mathematical and physical interpretation problems	186
Materials characterization and fabrication problems	186
Microelectrode array possibilities	187

Koichi Aoki 196
   Microcylinder Electrodes 196
      Chronoamperometry 197
      Linear sweep voltammetry 197
      Chronopotentiometry 197
   Normal pulse (NP) and differential pulse (DP)
   voltammetry 198
   Pulse voltammetric current-potential curves for
   electrode kinetics 198
   Microband electrodes 198
      Chronoamperometry 199
      Linear sweep voltammetry 199
      Electrode kinetics at microdisk electrodes . . 199
A. Bezegh and J. Janata 201
   Preparation of ultramicroelectrodes 201
   Temperature effects 201
   Warburg impedance of ultramicroelectrodes 201
John Bixler 210
Graham Cheek 211
Andrew Ewing 220
   Summary of fabrication, characterization and
   application of ultrasmall carbon-ring electrodes . 220

      Fabrication and characterization by electron
         microscopy 220
      Characterization by voltammetry 220
      Shielded carbon-ring microelectrodes 221
      Intracellular voltammetry with carbon-ring
         microelectrodes 223
Larry Faulkner 225
   Current multiplication 225
   Current-to-photon conversion 232
Stephen Feldberg 240
   Applications of microelectrodes 240
      Biological sensors 240
      Novel physical-chemical measurements 240
      Electroanalytical enhancement 240
   Theory 241
Martin Fleischmann 242
   Connections with other fields of research 242
   Connections with other areas of research in
   electrochemistry 245
      (i) Novel ways of investigating established
         systems 245
      (ii) Reinterpretation of the behavior of known
         systems 246
      (iii) Earlier investigations using
         microelectrodes 246
Jurgen Heinze 255
Charles Martin 263
Royce Murray 273
   Small size 273

```
        Arrays . . . . . . . . . . . . . . . . . . . . . 274
Keith Oldham . . . . . . . . . . . . . . . . . . . . . . 276
Janet Osteryoung . . . . . . . . . . . . . . . . . . . . 289
Stanley Pons . . . . . . . . . . . . . . . . . . . . . . 293
Hannah Reller . . . . . . . . . . . . . . . . . . . . . 297
Debra R. Rolison . . . . . . . . . . . . . . . . . . . . 298
Parbury P. Schmidt . . . . . . . . . . . . . . . . . . . 301
    Introduction . . . . . . . . . . . . . . . . . . . . 301
    The simulation of electron transfer to
        microelectrodes . . . . . . . . . . . . . . . . 302
Benjamin R. Sharifker . . . . . . . . . . . . . . . . . 316
    (a) Square lattice . . . . . . . . . . . . . . . . . 319
    (b) Hexagonal lattice . . . . . . . . . . . . . . . 322
    (c) Random array . . . . . . . . . . . . . . . . . . 324
Attilla Szabo . . . . . . . . . . . . . . . . . . . . . 326
    Hemicylinder, band, disk and ring electrodes . . . 326
R. Mark Wightman . . . . . . . . . . . . . . . . . . . . 333
Marek Wojciechowski . . . . . . . . . . . . . . . . . . 334
```

6
Selected Bibliographies   341

```
James L. Anderson . . . . . . . . . . . . . . . . . . . 341
Koichi Aoki . . . . . . . . . . . . . . . . . . . . . . 342
Alan M. Bond . . . . . . . . . . . . . . . . . . . . . . 344
Martin Fleischmann . . . . . . . . . . . . . . . . . . . 345
Royce W. Murray . . . . . . . . . . . . . . . . . . . . 347
Janet G. Osteryoung . . . . . . . . . . . . . . . . . . 349
Stanley Pons . . . . . . . . . . . . . . . . . . . . . . 351
Hannah Reller . . . . . . . . . . . . . . . . . . . . . 353
Attila Szabo . . . . . . . . . . . . . . . . . . . . . . 353
Dennis Tallman . . . . . . . . . . . . . . . . . . . . . 354
R. Mark Wightman . . . . . . . . . . . . . . . . . . . . 355
```

Index   **359**

PREFACE

When one of us, Stanley Pons, arrived at the University of Southampton in 1975 to complete graduate studies, he was amazed at the high level of sophistication that electrochemical experimentation had attained. For example, he noted that a certain Professor, one Martin Fleischmann, was clearing the corridors outside the laboratories in preparation for a high velocity extrusion experiment. One graduate student was about to launch a molten glass arrow from a crossbow. The molten glass contained a lead wire that would soon have its diameter rapidly decreased while having its length correspondingly increased. Pons was now absolutely sure that he had made the proper choice of graduate schools (well, almost). In any case, the devices that Fleischmann was preparing were ultramicroelectrodes. In a few short years, this type of device would prove to be the basis for new and exciting measurements in the electrochemical sciences. We are just at the beginning of this new area, and this volume is intended to provide workers with an insight to the possibilities that the area holds.

We gratefully acknowledge the Office of Naval Research and the National Science Foundation for support of many of us in this new area of electrochemistry. Program officers from those organizations, Dr. Jerry J. Smith and Henry N. Blount III, reviewed and recommended funding of an application for an extensive forum on the state of the art of electrochemical studies at ultramicroelectrodes which was subsequently held at the Homestead Lodge at Midway, Utah in January of 1986. The forum brought together an enthusiastic group of international scientists who were either established or who were beginning research efforts in the field. About 50 scientists attended the meeting. The organizers thank the participants for their part in the effort. We believe that their contributions gave us a deeper insight into the fundamental aspects of the field, and helped to define directions that the area would have to take to become a fruitful and successful area of science. It is clear from the attitude of the participants that the opportunity for a great variety of novel studies in electrochemistry is at hand; these include kinetic, thermodynamic, and electroanalytical measurements under conditions that heretofore have been impossible.

This volume contains the contributions to the Utah meeting on Ultramicroelectrodes from those who wished them published. We have included them as received, and have not had them critically reviewed. Shortly after the meeting, the organizers decided to add to the contributions a general review of selected research efforts in the field, theoretical considerations, a section on the fabrication of the devices, and some ideas on the future directions that might be taken in the area. We felt that a lesser effort would be rapidly outdated in light of the recent intensity of research in the area. The reviews are not meant to be exhaustive, but to provide a good cross section of the many applications that are possible in the area.

In addition, we decided to make the document available to the general scientific community by publishing it as a monograph. We thank Sheila Pons for her efforts in the organization, composition, and preparation of the finished pages of this volume. Appreciation is also expressed to Dr. Carol Korzeniewski and John Daschbach for reviewing parts of the manuscript.

List of Figures

Plot of the current parameter vs. overpotential as a
function of the rate parameter 31

Plot of current parameter vs. overpotential for a fast
electron transfer reaction 32

Plot of dimensionless concentration changes as a function
of radical position for a disk 33

Plots of dimensionless concentration changes as a function
of radial position for a ring 34

Plot of the square root of the dimensionless transition time
as a function of the inverse of the dimensionless flux . 45

Cole-cole plot of the dimensionless impedances 56

Equivalent circuit for a microelectrode 59

Schematic depiction of the use of polymer modified electrodes
for the determination of metal ions in solution. 159

Differential pulse voltammogram for a 25 micron diameter
platinum electrode modified with a $[Os(v-bpy)_3]^{+2}$/v-bpy 162

Forward (A); reverse (B); and difference (C) currents of a
square wave voltammogram for a platinum electrode modified
as in Figure 2 and after exposure to a $1 \times 10^{-6}\underline{M}$ solution of
Fe(II). 163

A. Structure of antipyrylazo III (AP-III)
  B. Structure of viologen polymer (PBV) 164

A. Cyclic voltammogram (cathodic sweep) for a platinum
electrode modified with PBV and AP-III.
  B. Differential pulse voltammogram for the electrode
in Figure 5A depicting the irreversible oxidation of AP-III
bound to the electrode. 165

Calibration curve for the determination of Ca^{+2} with an
electrode modified with PBV and AP-III. Open circle:
response for a $10^{-6}\underline{M}$ solution of calcium after addition of
Mg at the $1 \times 10^{-3}\underline{M}$ level. 167

Ohmic drop on the basis of primary current distribution at an ultramicroelectrode under planar diffusion, for three geometries of the electrode: spherical, disk and cylinder. 174

Model for chemically implanted electrode in brain tissue. . . 176

Experimental results for a chemically implanted electrode in brain tissue. Detection of ascorbate. 178

Conformal mapping of the diffusion field at a band electrode. 179

The high and low frequency response. 204

Equivalent circuit representation of the disk ultramicroelectrode. 205

Cole-Cole plot for ultramicroelectrodes. 206

Comparison of the effective diffusional area for the disk and ring ultramicroelectrodes. 208

Voltammogram of 1.9m$\underline{M}$ hydroquinone in benzene/0.30$\underline{M}$ THAP. Working electrode: 5 micrometer diameter platinum. Scan rate: 30mV/s. 212

Ratios of limiting currents (hydroquinone/ferrocene) as a function of added supporting electrolyte.
  1 Acetonitrile/TEAP: [H$_2$Q] = 0.64m$\underline{M}$ [Fc] = 0.59m$\underline{M}$
 M Methylene/TBAP: [H$_2$Q] = 0.91m$\underline{M}$ [Fc] = 0.86m$\underline{M}$
   chloride
 : Benzene/THAP: [H$_2$Q] = 1.90m$\underline{M}$ [Fc] = 1.74m$\underline{M}$. 214

Infrared spectra of hydroquinone in various solvent systems.
    (a) 5.2m$\underline{M}$ hydroquinone in acetonitrile (left spectrum)
  5.2m$\underline{M}$ hydroquinone in acetonitrile/0.10$\underline{M}$ TEAP (right spectrum)
  (b) 4.0m$\underline{M}$ hydroquinone in methylene chloride (left spectrum)
    4.0m$\underline{M}$ hydroquinone in methylene chloride/0.10$\underline{M}$ TBAP (right spectrum)
    (c)   10m$\underline{M}$ hydroquinone in benzene (left spectrum)
      10m$\underline{M}$ hydroquinone in benzene/0.15$\underline{M}$ THAP (right spectrum) 216

Ratios of limiting currents (quinone/ferrocene) as a function of added THAP concentration (benzene solutions).
  1 2.7m$\underline{M}$ 1,4-napthoquinone/2.0m$\underline{M}$ ferrocene
  M 2.7m$\underline{M}$ 2-hydroxy-1,4-napthoquinone/1.7m$\underline{M}$ ferrocene 218

Circuit diagram of the current multiplier. Reproduced from Reference 1. 227

Osteryoung square-wave voltammograms for (a) 0.1 mM and (b) 0.01 M ferrocene in acetonitrile with 0.1 M tetrabutylammonium fluoroborate. Amplitude, 25 mV; step height, 4 mV; freq., 15 Hz. Reproduced from Reference 1. . . 229

Instrument for square wave voltammetry in schematic form. AM1 and AM2 are AD534 analog multipliers. Logic circuitry (not shown) synthesizes the pulse trains at A, B, and C. 230

Barker square wave voltammograms for 0.1 mM ferrocene in acetonitrile containing 0.1 M tetrabutylammonium fluoborate. Distance between ticks is 600 mV. Curves are displaced horizontally for clarity. 233

Relative peak current vs. square root of frequency for ferrocene. System as in Figure 4. Top frequency is 50 KHz.   234

Schematic diagram of the current-to-photon converter. A bias is applied as indicated, so that the LED is emitting at a low level even at zero cell current. Pulses are applied from a fast pulse generator. 236

Data obtained with the apparatus of Figure 6 via time-correlated single-photon counting. (a) For an RC dummy with 1 microsecond time constant. Observed decay time was 1.2 microsecond. (b) Current transient in a cell with two Pt ultramicroelectrodes (25 micrometer diameter) in 0.2 M potassium sulfate.   238

Current-time transient for the deposition of a single droplet of mercury from a solution 0.2m$\underline{M}$ Hg$_2$(NO$_3$)$_2$ + 10m$\underline{M}$ HNO$_3$; h = 5mV; 4mm radius carbon disk electrode. 249

The deposition of two growth centers of a-PbO$_2$ onto an 8mm diameter C-microelectrode. Solution composition: 0.1$\underline{M}$ Pb(Ac)$_2$ + 1$\underline{M}$ HAc; h = 400mV; deposition time = 120s. . . 250

Current time transient for the deposition of a-PbO$_2$ onto an 8mm C-microelectrode. 251

Schematic diagram of the procedure used to prepare ultramicrodisk electrode arrays. 264

Electron micrograph of the surface of an ultramicroelectrode prepared from Nuclepore membrane with 1.0mm diameter pores. 266

Electron micrograph of Pt fibrils obtained after polycarbonate host membrane is dissolved away from the ultramicroelectrode shown in Figure 2. 267

Cyclic voltammograms at 5mV/s in 1m$\underline{M}$ ferrocene, 0.4$\underline{M}$ Bu$_4$NClO$_4$, acetonitrile for typical r=0.1mm and 0.5mm ultramicroelectrodes for a macro-sized Pt electrode (A=0.5cm^2). 269

Cyclic voltammograms at various scan rates for a typical r=0.5mm ultramicroelectrode in 1m$\underline{M}$ ferrocene, 0.4$\underline{M}$ Bu$_4$NClO$_4$, MeCN. 271

Microelectrode disadvantages. 277

Microelectrode advantages. 278

Types of microelectrodes. 279

Electrode dimensionality. 282

Characteristic lengths. 283

Resistance and dimensionality. 285

Ions in microelectrode cell. 287

List of Tables

Mass transport coefficient ratios for the disk and ring . . . 36

Values of the Φ_1 function 44

Values of the Φ_2 function 49

Values of the Φ_3 function 53

Values of the Φ_4 and Φ_5 functions 57

A Survey of Fabricated Ultramicroelectrodes 68

Epoxy Resins used in the Construction of
Ultramicroelectrodes 76

Chemical and Electrochemical Pretreatment Procedures for
Cylindrical Ultramicroelectrodes 82

Connection between the Investigation of Electrochemical
Systems using Microelectrodes and Other Fields of Research 244

PARTICIPANTS
at The Utah Conference on Ultramicroelectrodes

HECTOR ABRUNA
DEPARTMENT CHEMISTRY
CORNELL UNIVERSITY
ITHACA NY 14853

CHRISTIAN AMATORE
LABORITOIRE DE CHIMIE
ECOLE NORMALE SUPERIEURE
75231 PARIS CEDEX 5
FRANCE

JAMES ANDERSON
DEPARTMENT CHEMISTRY
UNIVERSITY OF GEORGIA
ATHENS GA 30602

KOICHI AOKI
GRADUATE SCHOOL-NAGATSUTA
TOKYO INSTITUTE OF TECHNOLOGY
YOKOHAMA 227
JAPAN

JAMES AUBURN
AT&T BELL LABS 6F-211
600 MOUNTAIN AVENUE
MURRAY HILL NJ 07974

ALAN BARD
DEPARTMENT CHEMISTRY
UNIVERSITY OF TEXAS AT AUSTIN
AUSTIN TX 787121

ANDRAS BEZEGH
DEPARTMENT OF BIOENGINEERING
UNIVERSITY OF UTAH
SALT LAKE CITY UT 84112

JOHN BIXLER
DEPARTMENT OF CHEMISTRY
STATE UNIVERSITY OF NEW YORK
BROCKPORT NY 14420

HENRY N. BLOUNT,III
NATIONAL SCIENCE FOUNDATION
WASHINGTON, DC 20550

JAMES BROPHY
DEPARTMENT OF PHYSICS
UNIVERSITY OF UTAH
SALT LAKE CITY UT 84112

STANLEY BRUCKENSTEIN
DEPARTMENT OF CHEMISTRY
STATE UNIVERSITY OF NEW YORK
BUFFALO NY 14214

GRAHAM CHEEK	U.S. NAVAL ACADEMY CHEMISTRY DEPARTMENT STOP 9B ANNAPOLIS MD 21402
MARK DEAKIN	DEPARTMENT OF CHEMISTRY INDIANA UNIVERSITY BLOOMINGTON IN 47405
ROYCE ENGSTROM	UNIVERSITY OF SOUTH DAKOTA DEPARTMENT CHEMISTRY VERMILLION SD 57069
ANDREW EWING	DEPARTMENT OF CHEMISTRY PENNSYLVANIA STATE UNIVERSITY UNIVERSITY PARK, PA 16802
EDWARD EYRING	DEPARTMENT OF CHEMISTRY UNIVERSITY OF UTAH SALT LAKE CITY UT 84112
LARRY FAULKNER	DEPARTMENT OF CHEMISTRY UNIVERSITY OF ILLINOIS URBANA IL 61801
STEPHEN FELDBERG	BROOKHAVEN NATIONAL LABS UPTON, NY 11973
ALANAH FITCH	DEPARTMENT OF CHEMISTRY UNIVERSITY OF WISCONSIN MADISON, WI 53706
MARTIN FLEISCHMANN	DEPARTMENT OF CHEMISTRY THE UNIVERSITY SOUTHAMPTON HAMPSHIRE ENGLAND SO9 5NH
JOHN FOLEY	DEPARTMENT OF CHEMISTRY UNIVERSITY OF UTAH SALT LAKE CITY UT 84112
JAMAL GHOROGHCHIAN	DEPARTMENT OF CHEMISTRY UNIVERSITY OF UTAH SALT LAKE CITY UT 84112
JOEL HARRIS	DEPARTMENT OF CHEMISTRY UNIVERSITY OF UTAH SALT LAKE CITY UT 84112
JED HARRISON	DEPARTMENT OF CHEMISTRY UNIVERSITY OF ALBERTA EDMONTON ALBERTA CANADA T6G 2G21

JURGEN HEINZE	INSTITUT FUR PHYSIKALISCHE CHEMIE ALBERTSTRASSE 21 D-7800 FRIEBERG GERMANY
JOHN HOWELL	BAS, INC. 2701 KENT AVE. WEST LAFAYETTE IN 47906
JIRI JANATA	DEPARTMENT OF BIOENGINEERING UNIVERSITY OF UTAH SALT LAKE CITY UT 84112
MARCIN MAJDA	DEPARTMENT OF CHEMISTRY UNIVERSITY OF CALIFORNIA BERKELEY CA 94720
CHARLES MARTIN	DEPARTMENT OF CHEMISTRY TEXAS A&M UNIVERSITY COLLEGE STATION TX 77843
RICHARD MCCREERY	DEPARTMENT OF CHEMISTRY THE OHIO STATE UNIVERSITY COLUMBUS, OHIO 43210
BARRY MILLER	AT&T BELL LABS ROOM 1D-350 600 MOUNTAIN AVE MURRAY HILL NJ 07974
MICHAEL GRATZEL	INSTITUT DE CHEMIE PHYSIQUE ECOLE POLYTECHNIQUE FEDERALE 1015-LAUSANNE SWITZERLAND
ROYCE MURRAY	DEPARTMENT OF CHEMISTRY UNIVERSITY OF NORTH CAROLINA CHAPEL HILL NC 27514
ROBERT NOWAK	OFFICE OF NAVAL RESEARCH 800 NORTH QUINCY AVENUE ARLINGTON VA 22217
KEITH OLDHAM	DEPARTMENT OF CHEMISTRY TRENT UNIVERSITY PETERBOROUGH ONTARIO CANADA K9J7B8
JANET OSTERYOUNG	DEPARTMENT OF CHEMISTRY STATE UNIVERSITY OF NEW YORK BUFFALO, NY 14214

STANLEY PONS   DEPARTMENT OF CHEMISTRY
                                   UNIVERSITY OF UTAH
                                   SALT LAKE CITY UT 84112

HANNAH RELLER   DEPARTMENT OF CHEMISTRY
                                   TEL AVIV UNIVERSITY
                                   TEL AVIV 69978 ISRAEL

DEBRA ROLISON   NAVAL RESEARCH LABORATORY
                                   CODE 6171
                                   WASHINGTON DC 20375

FERESHTEH SARFARAZI   DEPARTMENT OF CHEMISTRY
                                   UNIVERSITY OF UTAH
                                   SALT LAKE CITY UT 84112

PARBURY SCHMIDT   DEPARTMENT OF CHEMISTRY
                                   OAKLAND UNIVERSITY
                                   ROCHESTER MI 48063

JERRY J. SMITH   NAVAL WEAPONS CENTER
                                   CHINA LAKE CA 93555

ATILLA SZABO   CHEMICAL PHYSICS DIVISION
                                   BLDG 2 ROOM B1-28
                                   NATIONAL INSTITUTES OF HEALTH
                                   BETHESDA MD 20205

DENNIS TALLMAN   DEPARTMENT OF CHEMISTRY
                                   NORTH DAKOTA STATE UNIVERSITY
                                   FARGO ND 58105

WOLFGANG THORMANN   CENTER FOR SEPARATION SCIENCE
                                   UNIVERISTY OF ARIZONA
                                   TUCSON AZ 85721

VICTORIA VICENTE-BENNETT   DEPARTMENT OF CHEMISTRY
                                   UNIVERSITY OF UTAH
                                   SALT LAKE CITY UT 84112

MARK WIGHTMAN   DEPARTMENT OF CHEMISTRY
                                   INDIANA UNIVERSITY
                                   BLOOMINGTON IN 47405

MAREK WOJCIECHOWSKI   DEPARTMENT OF CHEMISTRY
                                   UNIVERSITY OF MARYLAND
                                   CANTONSVILLE MD 21228

1

Introduction and Overview

NEW STUDIES WITH ULTRAMICROELECTRODES

The first major research programs involving ultramicroelectrodes were carried out by Martin Fleischmann and co-workers at the University of Southampton in the late 1960's. Many of the early data appeared in the M. Phil. thesis of H.Y.S. Lui in 1975 and the Ph.D. thesis of David Swan. Other workers began utilizing the devices in the early 1980's and demonstrated their utility in electroanalytical and bioelectrochemical studies (e.g., Wightman, Osteryoung, and others). Most of this work is discussed elsewhere in this volume.

Reducing the area of a single electrode has three major consequences: i) Mass transport rates to and from the electrode are increased; ii) The double layer capacitance is reduced due to the decrease in surface area; iii) Ohmic losses (the product of the electrode current and the solution resistance) are reduced due to the diminished current. (Other consequences include miniaturized cells, decreased material costs, and reduced electronic instrumentation requirements.) With very small electrodes, new experiments can be performed in environments which were previously inaccessible.

There are dramatic changes to "conventional" electrochemical responses when very small electrodes are substituted. These are usually due to a reduction in the transient diffusion response, with a tendency toward steady state mass transport. The conventional cyclic voltammogram therefore is transformed to a "polarographic-type wave" which is easier to treat analytically. The technique of chronopotentiometry, the response of which is usually considerably distorted in practice due to capacitative effects now promises to be revived as a more useful analytical technique with new applications (*vide infra*). In chrono-amperometry, transient measurements at ultramicroelectrodes for determining fast electrochemical reactions can be made at higher perturbation speeds with less capacitative current effects.

In addition, ultramicroelectrodes offer the advantage of being able to investigate electrochemistry in highly resistive media. Experiments can be conducted in solutions which contain no added supporting electrolyte, in frozen solutions, and in the gas phase; these media provide new grounds of investigation in physical and analytical chemistry.

The following pages will describe some of the areas of research open to electrochemists through the use of ultramicroelectrodes. The authors present a synopsis of past work in these areas, and some thought as to what are likely to be fruitful areas of future research. This chapter is divided into five sections describing electrochemistry in: I-The Liquid Solution Phase; II-The Solid Solution Phase; III-The Gas Phase; IV-Theoretical Considerations, and V-New Constructions and Systems.

STUDIES IN THE LIQUID SOLUTION PHASE

Studies of fast electrode reactions and of fast homogeneous reactions
Present status

As will be seen later, the advantage of rapid rates of mass transport to ultramicroelectrodes makes it possible to measure heterogeneous and homogeneous kinetics under essentially steady state conditions.

It has been shown (1) that the kinetics of fast reactions in solution coupled to electrode reactions (such as the chemical steps in e-c-e or displ reaction sequences) can be evaluated by making steady state measurements as a function of electrode size. These measurements allow new diagnostic criteria to be established for distinguishing between different reaction schemes. Furthermore, heterogeneous rate constants up to a few tenths of cm-s^{-1} have been measured under steady state conditions at small ring electrodes (2).

The advantages of steady state measurements cannot be overemphasized in solution phase electrochemistry, especially in cases where fast kinetic parameters are sought. Whereas a perturbative technique for kinetic study requires disturbing a system from equilibrium and measuring the response as it relaxes to equilibrium, steady state techniques do not. The measured steady state response is directly related to the kinetics, and is not generally complicated by mass transport or migration. Thermodynamic and kinetic data may be obtained simultaneously.

Future considerations

The rates of mass transport increase rapidly with the decreasing size of the ultramicroelectrode (1). For the study of fast electrode kinetics in the steady state, it is therefore desirable to develop smaller electrode structures. With the present technology, it is feasible to fabricate ultramicro-electrodes with dimensions on the order of hundreds of angstroms. Using electrodes of these dimensions, it should be possible to observe, under steady state conditions, electron transfer reaction rate constants greater than 50cm-s^{-1}.

The transition time to the steady state is on the order of

$r^2/4D$ where r is the radius of the ultramicroelectrode and D is the diffusion coefficient. For electrodes having dimensions of 100 angstroms this is about 25 nanoseconds. For first order reactions, this timescale defines an upper limit for reaction rate constant measurements to be near $10^7 s^{-1}$. Such measurements have been virtually impossible using existing electrochemical techniques due to charging effects and mass transport limitations.

The electrochemical study of reactions in nonconducting liquid solutions; catalysis
Future considerations

Comparison of kinetic and thermodynamic electrochemical data with those from other well established techniques, such as ESR, NMR and others, has been difficult since electrochemical experiments always require the presence of an electrolyte. The electrolyte invariably will complicate the interpretation of the data, due to its ability to ion-pair the reactants, intermediates, or products. Studies of the mechanism of heterogeneous electron transfer reactions will thus be most effectively made in the absence of electrolytes. Such measurements are possible at ultramicroelectrodes.

Electrocatalytic reactions (except at dispersions (*vide supra*) have not yet been studied at ultramicroelectrodes. There are many important systems to be studied under these conditions; they include hydrogen evolution reactions, oxygenation and hydrogenation of organic fuels, hydride reactions, and the generation of strong bases, to mention a few.

Dispersions of ultramicroelectrodes: the study of nonelectrochemical catalysis
Present status

Dispersions of ultramicroelectrodes offer an opportunity to

study catalytic reactions directly. A novel feature of the technique is that it is not necessary to make direct electronic contact between the ultramicroelectrodes and the external circuit. At the same time, it is possible to produce large total currents in the cell, so that the system has practical uses. Results of experiments on the hydrogen evolution/ oxidation and oxygen evolution/reduction reactions at dispersions of ultramicroelectrodes suspended between feeder electrode plates at high potentials have been reported (4). The high field gradient in the cell drives the electrochemical reactions on alternate halves of the electrode particle i.e., each particle becomes a bipolar cell. It was pointed out that it is possible to study electrode processes in poorly conducting or nonconducting media. The results obtained for the hydrogen/oxygen system fit predictions derived from a simple model based on the concepts of the mixed electrode potential with mass transfer effects considered. Catalyst supported dispersions of oxides and zeolites (5) have also been investigated.

Future considerations

The conditions used for the experiments reported in the previous work were close to those of conventional heterogeneous liquid phase catalyses; it is therefore feasible that conventional catalytic processes can be investigated (and monitored) provided that electron transfer is involved somewhere in the overall mechanism. As measurements can be made in nonconducting media it is evident that such investigations could be extended to semiconductor (and possibly insulator) catalysts. Investigations of catalytic systems will require extensions of the modelling that has been made to date. These must include the effects of reactions and concentration changes in solution.

The results obtained also show that it should be possible to modify catalytic rates by means of externally applied fields.

Dispersions of ultramicroelectrodes: the study and implementation of efficient electrochemical syntheses
Present status

Electrochemical synthesis has been the subject of study of a great many researchers for many years. Although there are only a few major organic electrosynthetic plants, there are many smaller specialty synthetic operations in production. The first experiments utilizing ultramicroelectrode synthetic cells have been performed in order to investigate their utility for practical synthesis (6). That work demonstrated the feasibility of driving hydrogenation reactions, functional activation of saturated fuels like methane to acetamide, the oxidation of benzene, and several other energetically difficult reactions in nonconductive media.

The extension of these studies to include reactions at electrode systems composed of supported metals on other substrates, such as zeolites and aluminas, have also been started (6), and are important because of the extra parameter which can be controlled in the system: ion exchange and controlled access to the reaction site.

The extension of the method to include semiconductor particles has also been undertaken (7). The evolution of hydrogen at dispersions of titanium dioxide is markedly enhanced over the bulk semiconductor analog. Reaction of a variety of anions, such as cyanide and sulphate, is two orders of magnitude faster than under similar conditions for the reactions at bulk semiconductors. The same reactions at illuminated dispersion systems are markedly faster than on planar electrodes. Decomposition of the particles during electrolysis by photocorrosion is inhibited due to the driving of the conjugate process at the other side of the semiconductor particle. Rapid rotation of the particle in the field assures rapid protection of the corroding surface after the catalytic charge transfer reaction has taken place.

Future considerations
The study of reaction mechanisms at semiconductor dispersions will be important. The implications for efficient photocatalytic reactions are that one can use both optical radiation and the feeder electrode potential to separate electron-hole pairs and increase the rate of mass transport. Other workers have demonstrated the efficiency of the photocatalytic decomposition of water at very small semiconductor particles such as titanium dioxide; particles where the size is such that the "band gap" (or at any rate the separation of filled and empty states) becomes affected by quantum size effects.

STUDIES IN THE SOLID SOLUTION PHASE

Fundamental electrochemical studies in the solid solution phase
Present status
It is feasible to perform electrochemical studies in solid eutectics, glasses and other solid solutions (8). Diffusion controlled waves are observed in a variety of solvent systems and temperatures. The currents, as expected, are diminished from those measured in the liquid phase. Electrochemical reactions have been studied in these solid solutions with and without added electrolyte.

Future considerations

A study of charge transport mechanisms in solids has a wide range of applications. Thermodynamic and kinetic parameters can be obtained in materials such as semiconductors and polymers. Temperature labile species such as free radicals and other reaction intermediates can be stabilized in low temperature matrices; electrochemical characterization of these species *in situ* is then possible.

Membrane coated ultramicroelectrodes
Present status

Several workers in electrochemistry and electroanalytical chemistry (e.g., Murray et al., Abruña, *vide infra*) have quickly realized the advantages in studying polymer modified electrodes mounted on ultramicroelectrode substrates; most of these experiments have been in the areas of miniaturization of electrochemical sensors and detectors; some have used the advantage of low ohmic resistance at ultramicroelectrodes to perform sensor/detector tasks in media of high specific resistivity.

Future considerations

While these applications are important, development of highly specific detectors and sensors will depend upon the specificity of the polymer substrate. The mechanisms of ion transport and electron transfer in solid materials are readily studied with ultramicroelectrodes. These studies are enhanced by using ultramicroelectrodes as substrate mounts since (a) the studies can be carried out over a wide range of electrolyte concentrations, even with no purposely added supporting electrolyte, (b) mass transport rates are accelerated due to spherical symmetry, and (c) fast steady states of mass transport

are established at the ultramicroelectrodes. Each of these
advantages is not attainable at large electrodes, and each
affects the parameters that must be studied in order to properly
evaluate mass transport models in membranes.

THE GAS PHASE

Gas phase electrochemistry and the study of reaction kinetics
Present status

It has been demonstrated that it is possible to electrochemically detect species in the gas phase under certain conditions
(9). Electrode assemblies are readily constructed that are
separated by surface conductive materials that allow migration/
diffusion of protons. The high mobility of the proton leads to
conditions that make it possible to observe electron transfer to
species that are incident on the ultramicroelectrode in the gas
phase. It is also possible to record potential-current plots by
the technique.

Future considerations

Many practical and fundamental questions regarding electron
transfer reactions lie unanswered. Voltammetric studies of
electrochemical reactions in the gas phase provides insight into
many of these problems. In the gas phase, the effects of
solvation will be reduced, and the reorganizational energies
will be more easily determined. The behavior of electroactive
species under these less ambiguous circumstances in turn provide
information regarding the contributions of inner and outer
sphere solvation reorganization in solution phase electrode
reactions. Also, reactions of a wider range of species can be

studied since the solvent and electrolyte which limit the potential window to low values are not present. Further, studies at low pressure are important in several areas of chemistry and are possible at ultramicroelectrodes.

Catalysis at ultramicroelectrodes
Future considerations

The questions to be addressed in this area are those concerned with the viability of monitoring gas phase catalytic reactions. The wide variety of gas phase catalytic reactions is well known; these include a wide range of hydrogenation reactions and specific oxidation reactions. One question is whether one can indeed measure external cell currents in the presence of a surface gas phase catalytic reactions.

Other reactions of interest include that of ethylene with oxygen at a silver electrode, the reactions of CO_2 at carbon electrodes, the oxidation of a variety of small molecule oxygen containing fuels at platinum, palladium, and rhodium ultramicroelectrodes, and organic oxidation and hydrogenation reactions.

Steam systems
Present status

One of the most important practical and fundamental systems to investigate in the gas phase is that of steam (10). Reactions of importance in steam include those at the surfaces of the container (hydrogen evolution, oxygen reduction, the oxidation of organics dissolved in the steam, metal dissolution and oxide formation) and reactions and adsorption of corrosion inhibitors and other additives. One basic goal for these studies is the monitoring of the quality of the steam for various practical applications.

Future considerations

The hydrogen evolution reaction in steam will be studied for fundamental interest. The reactions of other species can be studied as a function of the partial pressures of the reactants and temperature. Studies in this medium necessitate the design of new types of corrosion resistant cells and electrodes, and their operation at elevated temperatures.

THEORETICAL CONSIDERATIONS

Investigation of electrochemical kinetics
Present status

There has been much activity in the development of the mathematical description of mass transport and electrochemical kinetics at ultramicroelectrodes. The analysis for common configurations such as the disk, sphere, ring, and line electrodes has been investigated for a variety of electrochemical experiments. (See Chapters 2 and 4.) While there has been some success in modelling these processes, especially in deriving steady state and time-dependent solutions, much work remains to be accomplished on coupled chemical reactions. Numerical and asymptotic solutions have been obtained for some cases, while the exact solution for the disk under various circumstances is presented in the theory chapter.

Future considerations

Solutions to kinetic problems are strongly affected by secondary and tertiary current distributions across the electrode surface. These effects are particularly significant at nonspherical ultramicroelectrode geometries since there can be large differences in mass transport coefficients across the surface; appropriate models must therefore contain a treatment of these effects. For the same reasons these treatments must include the changes in the current distributions as a function of potential.

In addition, there are a variety of new structures that will require analysis. One example is that of the ring-ring system. This configuration consists of two (or more) concentric ring ultramicroelectrodes. This configuration allows experiments similar to those of the mechanically rotated ring disk electrode systems, except at increased mass transport rates, and in highly resistive media. The time dependent solutions for this problem are also achievable by use of the analytical models presented in the theoretical chapter.

Theoretical treatment of the electrical double layer: investigations with the use of ultramicroelectrodes

Future considerations

Theoretical models of the electrical double layer must be generalized to include effects of dilute electrolyte solutions on electrodes when their dimensions become comparable to the Debye length. The double layer behavior will become dependent on electrode size and shape (e.g., sphere, disk, ring or line). One can envisage conditions where the charge density on ring or line electrodes becomes sufficiently small that the system becomes effectively two- rather than three-dimensional. Microelectrodes therefore become novel probes of double layer structure, and the double layer effects on electrochemical reactions become dependent on electrode geometry. Furthermore, the coupling between the fluxes and double layer structure will be dependent on electrode geometry.

MISCELLANEOUS APPLICATIONS AND PROBLEMS FOR INVESTIGATION. NEW CONSTRUCTIONS AND NOVEL SYSTEMS

Sheets of imbedded ultramicroelectrodes

Present status

Ultramicroelectrode arrays are useful in a variety of applications (see Chapters 3 and 4). Arrays can be conveniently prepared by several methods, such as (a) the coating of metals (gold, platinum, palladium) in the channels and on one surface of microchannel detector plate substrates to form exposed disks or rings at the opposite face; (b) the use of photolithographic techniques to form arrays of ultramicroelectrodes on silicon substrates; (c) the embedding of small conducting fibers (e.g., carbon or metal fibers) in an insulating substrate; (d) the repetitive drawing of large numbers of metal filled capillary bundles, similar to the drawing of capillary light pipes; and (e) the use of micropore filters and similar materials having liquid metals forced through from the back surface plane to expose arrays of metallic ultramicroelectrodes.

Future considerations

There is a need prescribed by synthetic and other practical applications to develop large arrays of ultramicroelectrodes. At these devices, large quantities of materials may be electrolyzed while at the same time retaining the advantages (low iR drop and reduced double layer capacitance) of ultramicroelectrodes. In addition, these arrays, if operated in a multiplexed data acquisition system, can be used as area detectors in the liquid, gas, or solid phase. Also these devices can be designed to be constructed of more than one type of specific sensor/detector, so that multicomponent analyses under a variety of ambient conditions might be carried out.

In energy producing electrochemical systems, the elimination of electrolytes makes it possible to test systems of higher free energy, such as alkali metal/ liquid hydrocarbon based electrolytic cells, as well as high energy fuel cell systems.

Statistical studies of stochastic processes, such as of nucleation and electrocrystallization can be enhanced by the use of multiplexed arrays of fixed geometry ultramicroelectrodes, since the time evolution of the processes on new electrodes may be examined without the repeated preparation of electrode surfaces.

The fabrication and coating of very small dimensioned disk and ring ultramicroelectrodes

Present status

It is possible with commercially available equipment to produce glass capillaries with inner diameters on the order of 1000 Å for intercellular biological studies. The equipment can be used to pull these capillaries in a linear distance of only a few microns, so that the length of the capillary is not so long as to prohibit the forcing of liquids through them. One is able to force metal screen printing inks through these capillaries which are then thermally reduced to the free metal (10).
In addition, it is possible to electrolytically and chemically etch fine metal wires to fine points, and imbed them in glass. This is accomplished by drawing the sharpened fibers through a small bead of glass solder under a microscope. The glass is caused to flow along the wire by controlled heating of the bead in a vertical stage oven at a controlled rate. Electrodes with tip dimensions on the order of 100 Å can be prepared by this technique (10).

The design of experiments with the dropping and sessile drop mercury ultramicroelectrode
Present status

Electrochemistry at the mercury interface is well characterized; it approaches the "ideal" polarized interface in aqueous solutions. In addition, very small spherical, or hemispherical (in an insulating plane) electrodes will exhibit the greatest rates of mass transport. It is therefore desirable to develop mercury ultramicroelectrodes. Several workers (*vide infra*) have developed mercury droplets on other substrates by nucleation and growth methods. A new type of mercury ultramicroelectrode has been developed, however, that is quite similar to the conventional dropping mercury electrode, except that the droplets are considerably smaller (11). The device consists of a capillary tube which contains a mercury reservoir bulb at one end, and a very small tip (0.8 to 7.0μm) at the other. The device is mounted in a controlled temperature oven and the mercury droplets are driven out thermally. Experiments have been performed which allow the observation of the transition from kinetic to mass transport control due to the rapid change in mass transfer coefficient as the droplet increases in size.

References

1. M. Fleischmann and S. Pons, J. Electroanal. Chem., 222 (1987) 107.
2. M. Fleischmann, Lasserre, J. Robinson and D. Swan, J. Electroanal. Chem., 177 (1984) 97.
   M. Fleischmann, Lasserre and J. Robinson, J. Electroanal. Chem., 171 (1984) 115.
3. J. W. Bixler, A. M. Bond, D. A. Lay, M. Fleischmann and S. Pons, Anal. Chim. Acta, 187 (1986) 67.
4. M. Fleischmann, J. Ghoroghchian and S. Pons, J. Phys. Chem., 89, (1985) 5530.
5. M. Fleischmann, J. Ghoroghchian, D.R. Rolison and S. Pons, J. Phys. Chem., 90 (1986) 6392.
6. J. Ghoroghchian, M. Fleischmann and S. Pons, in preparation.
7. S. Pons and M. Fleischmann, in preparation.
8. A. M. Bond, M. Fleischmann and J. Robinson, J. Electroanal. Chem. 180 (1984) 257.
9. J. Ghoroghchian, F. Sarfarazi, T. Dibble, J. Cassidy, J. J. Smith, A. Russell, M. Fleischmann and S. Pons, Anal. Chem., 58 (1986) 2278.
10. R. Brina, S. Pons and M. Fleischmann, J. Electroanal. Chem., submitted.
11. M. Fleischmann, S. Pons, J.W. Pons and J. Daschbach, J. Electroanal. Chem., submitted.

2

Selected Theoretical Topics

Electrochemical behavior of disks and rings

It is well known that closed form solutions (or, at any event, simple solutions) can be obtained for most types of electrochemical experiments using hemispherical or spherical microelectrodes provided that the reaction kinetic terms are first order or pseudo-first order. Unfortunately the applicability of such types of microelectrodes has so far been somewhat restricted e.g., to the electrodeposition of ensembles (1,2) or single mercury droplets (3,5) or the electrolysis of dispersions (6,7). Disk, and the recently introduced ring microelectrodes (4,8,9), are more easily constructed (see Chapter 3) but the necessary mathematical analysis has so far proved to be rather intractable.

The mathematical difficulties of analyzing problems in the cylindrical coordinate system are due to the discontinuities at the edges of the electrodes where the diffusion limited flux for a reversible system becomes infinite (the combined effects of the finite rates of the surface reactions and of the distribution of potential and concentration i.e., the "tertiary current distribution," however, will limit the rates at the edges for real systems). A variety of analytical and simulation procedures have been used in attempts to develop adequate descriptions of mass transfer and of the chronoamperometric and chronopotentiometric responses of disk and ring electrodes, e.g., see (9-23); some of these approaches will be described elsewhere in this volume. Disk and ring electrodes, however, have the advantage that quasi-spherical diffusion fields are established at relatively short times[1]. These spherical diffusion fields give high rates of mass transfer to the surfaces of the microelectrodes so that the kinetics of fast electrode reactions and of fast reactions in solution can be studied under *steady state* conditions e.g., see (28). It is therefore always useful to examine whether the steady state behavior can be predicted directly and in this chapter we first of all outline a simple approach based on the properties of discontinuous integrals (see e.g., (29-31)) which allows the prediction of the mass transfer coefficients for constant concentration and constant flux conditions. We illustrate how these coefficients can be used to analyze polarization curves and also discuss the application of more approximate methods to the analysis of electrochemical experiments. In the final section we discuss the extension of the approach to predict the nonsteady state behavior of disk electrodes for a variety of electrochemical experiments.

General

In the steady state the behavior is described by diffusion using the cylindrical coordinate system

$$\frac{\partial^2 c}{\partial r^2} + \frac{1}{r}\frac{\partial c}{\partial r} + \frac{\partial^2 c}{\partial z^2} = 0 \qquad [1]$$

and we investigate the possibility of this equation having a

---

[1] By contrast the rate of mass transfer to line or band electrodes varies with $(\ln t)^{-1}$ at long times (24-27).

solution for the appropriate boundary conditions. We first separate the variables

$$C = R(r)\, Z(z) \tag{2}$$

and rewrite [1] as

$$Z(z)\frac{d^2R}{dr^2} + \frac{Z(z)}{r}\frac{dR}{dr} + R(r)\frac{d^2Z}{dz^2} = 0 \tag{3}$$

Equation [3] will hold provided

$$\frac{d^2R}{dr^2} + \frac{1}{r}\frac{dR}{dr} + \lambda^2 R = 0 \tag{4}$$

and

$$\frac{d^2Z}{dz^2} + \lambda^2 Z = 0 \tag{5}$$

where λ is to be determined by the appropriate model for the experiment. Equations [4] and [5] have the solutions

$$R(r) = AJ_0(\lambda r) + BY_0(\lambda r) \tag{6}$$

and

$$Z(z) = \exp(-\lambda Z) \tag{7}$$

where J_0 and Y_0 are Bessel functions of the first and second kind of order zero. Since R must remain finite as $r \to 0$, we have

$$B = 0 \qquad [8]$$

We thus predict a general solution of the form

$$C = \sum_{k=0}^{\infty} a_k \exp(-\lambda_k z) J_0(\lambda_k r) \qquad [9]$$

The a_k and λ_k are sought so that [9] satisfies the appropriate boundary conditions for any particular electrode geometry. However, the λ_k are not determinable as the roots of any conceivable transcendental equation and we therefore conclude that λ must vary continuously in $0 < \lambda < \infty$. We therefore replace [9] by

$$C = C^{\infty} - \int_0^{\infty} f(\lambda) \exp(-\lambda z) J_0(\lambda r) \, d\lambda \qquad [10]$$

which will be the solution of [1] provided $f(\lambda)$ is chosen to satisfy the boundary conditions.

The disk

For a reversible reaction at the surface of a microdisk electrode we will have

$$C = C^s = C^\infty - \Delta C, \quad 0 < r < a, \quad z = 0 \qquad [11]$$

$$\frac{\partial C}{\partial z} = 0, \quad\quad r > a, \quad z = 0 \qquad [12]$$

where a is the radius of the disk, C^s is the surface concentration, and ΔC is the difference between the bulk and surface concentration. The mixed boundary conditions [11] and [12] show that the integral [10] must be discontinuous at $r = a$ and that we must have in addition

$$\left(\frac{\partial C}{\partial z}\right)_{z=0} = \int_0^\infty \lambda\, f(\lambda)\, J_0(\lambda r)\, d\lambda = 0 \qquad [13]$$

for $r > a$ while for $r < a$ the gradient is arbitrary except that it cannot be zero.

The discontinuous integrals

$$\int_0^\infty \sin(\lambda a) J_0(\lambda r)\frac{d\lambda}{\lambda} \quad \begin{cases} \dfrac{\pi}{2}, & r < a \qquad [14] \\ \sin^{-1}\left(\dfrac{a}{r}\right), & r > a \qquad [15] \end{cases}$$

associated with

$$\int_0^\infty \sin(\lambda a) J_0(\lambda t) d\lambda \quad \begin{cases} \dfrac{1}{(a^2 - r^2)^{1/2}}, & r < a \qquad [16] \\ 0, & r > a \qquad [17] \end{cases}$$

satisfy [11] and [12] respectively provided

$$f(\lambda) = \frac{2}{\pi} \Delta C \sin(\lambda a) \qquad [18]$$

i.e., we obtain the solution

$$C = C^\infty - \frac{2}{\pi}(C^\infty - C^s) \int_0^\infty \exp(-\lambda z) \sin(\lambda a) J_0(\lambda r) \frac{d\lambda}{\lambda} \qquad [19]$$

We are interested in the total flux to the surface of the disk

$$F = 2\pi D \int_0^a \left(\frac{\partial C}{\partial z}\right)_{z=0} r\, dr$$

$$= 4D (C^\infty - C^s) \int_0^a \int_0^\infty \sin(\lambda a) J_0(\lambda r) r\, dr\, d\lambda$$

$$= 4D (C^\infty - C^s) a \int_0^\infty \sin(\lambda a) J_1(\lambda a) \frac{d\lambda}{\lambda}$$

$$= 4D (C^\infty - C^s) a \qquad [20]$$

since

$$\int_0^a \lambda r J_0(\lambda r)\, dr = a J_1(\lambda a) \qquad [21]$$

and

$$\int_0^\infty \sin(\lambda a) J_1(\lambda a) \frac{d\lambda}{\lambda} = 1 \qquad [22]$$

Equation [20] shows that the mass transfer coefficient for

constant concentration conditions at the surface of a microdisk

$$(k_m)_c = \frac{4D}{\pi a} \qquad [23]$$

The result [20] has been known for some time from analagous analyses of problems in electrostatics (32) and heat transfer (30,31). Indeed Lord Kelvin referred to an experimental determination by Cavendish of the ratio of the capacitance of a sphere to that of a disk which would give the result [23].

For irreversible reactions we will have a constant flux Q (mol cm^{-2} s^{-1}) to the surface of the disk (i.e., a sink of strength Q) and replace [11] by

$$D\left[\frac{\partial c}{\partial z}\right] = Q \quad, \quad 0 < r < a \quad, \quad z = 0 \qquad [24]$$

We now make use of the discontinuous integrals

$$\int_0^\infty J_0(\lambda r) \, J_1(\lambda a) \, d\lambda = \begin{cases} 0, & r>a \qquad [25] \\ \frac{1}{2a}, & r=a \qquad [26] \\ 1/a, & r<a \qquad [27] \end{cases}$$

and obtain the concentration distribution

$$c = c^\infty - \frac{Qa}{D} \int_0^\infty \exp(-\lambda z) \, J_0(\lambda r) \, J_1(\lambda a) \, \frac{d\lambda}{\lambda} \qquad [28]$$

It is convenient to define a mass transfer coefficient for constant flux conditions, $(k_m)_Q$ (cm s^{-1}), by setting the average concentration at the surface of the disk equal to zero. We obtain

$$C_{Av} = C^\infty - \frac{2Qa}{Da} \int_0^a \int_0^\infty J_0(\lambda r) J_1(\lambda a) \frac{r}{\lambda} \, dr \, d\lambda$$

$$= C^\infty - \frac{2Qa}{D} \int_0^\infty \left[J_1(\lambda a) \right] \frac{d\lambda}{\lambda^2}$$

$$= C^\infty - \frac{8Qa}{3\pi D} \qquad [29]$$

We obtain the mass transfer coefficient

$$(k_m)_Q = \frac{3\pi D}{8a} \qquad [30]$$

a result which has also been known for some time e.g., see (31).

The ring

A straightforward result is obtained for constant flux into the ring. The concentration distribution due to a disk source of strength Q and radius a is given by [28] but with a positive sign for the source. We now superpose a disk sink of strength Q and radius b on the disk of radius a so that the combined source and sink give a ring sink of strength Q and dimension a < r < b. The flux distribution is

$$F = \begin{cases} 0, & r<a \\ -Q, & a<r<b \\ 0, & r>b \end{cases} \qquad [31]$$

and the concentration distribution is

$$C = C^\infty + \frac{Qa}{D}\int_0^\infty \exp(-\lambda z)\, J_0(\lambda r)\, J_1(\lambda a)\, \frac{d\lambda}{\lambda}$$

$$- \frac{Qb}{D}\int_0^\infty \exp(-\lambda z)\, J_0(\lambda r)\, J_1(\lambda b)\, \frac{d\lambda}{\lambda} \qquad [32]$$

It is convenient to express the integrals in [32] in terms of the relevant hypergeometric functions. We are especially interested in the concentration distribution over $z = 0$ as well as the average concentration over the surface of the ring, C_{Av}. We obtain

$$C^s = C^\infty + \frac{Qa}{D}\int_0^\infty J_0(\lambda r)\, J_1(\lambda a)\, \frac{d\lambda}{\lambda}$$

$$- \frac{Qb}{D}\int_0^\infty J_0(\lambda r)\, J_1(\lambda b)\, \frac{d\lambda}{\lambda}$$

$$= C^\infty + \frac{Qa}{D}\, {}_2F_1\!\left\{\tfrac{1}{2}, -\tfrac{1}{2};\, 1;\, \frac{r^2}{b^2}\cdot\frac{b^2}{a^2}\right\} - \frac{Qb}{D}\, {}_2F_1\!\left\{\tfrac{1}{2}, -\tfrac{1}{2};\, 1;\, \frac{r^2}{b^2}\right\}$$

$$\text{for } r<a$$

$$= C^\infty + \frac{Qa^2}{2Dr}\, {}_2F_1\!\left\{\tfrac{1}{2}, \tfrac{1}{2};\, 2;\, \frac{b^2}{r^2}\cdot\frac{a^2}{b^2}\right\} - \frac{Qb}{D}\, {}_2F_1\!\left\{\tfrac{1}{2}, -\tfrac{1}{2};\, 1;\, \frac{r^2}{b^2}\right\}$$

$$\text{for } a<r<b$$

$$= C^\infty + \frac{Qa^2}{2Dr}\, {}_2F_1\!\left\{\tfrac{1}{2}, \tfrac{1}{2};\, 2;\, \frac{b^2}{r^2}\cdot\frac{a^2}{b^2}\right\} - \frac{Qb^2}{2Dr}\, {}_2F_1\!\left\{\tfrac{1}{2}, \tfrac{1}{2};\, 2;\, \frac{b^2}{r^2}\right\}$$

$$\text{for } r>b \qquad [33]$$

We can express [33] as a dimensionless concentration variable

$$C_{z=0} = \frac{(C^\infty - C)D}{Qb}$$

$$= {}_2F_1\left\{\frac{1}{2}, -\frac{1}{2}; 1; \frac{r^2}{b^2}\right\} - \frac{a}{b} {}_2F_1\left\{\frac{1}{2}, -\frac{1}{2}; 1; \frac{r^2}{b^2} \cdot \frac{b^2}{a^2}\right\}$$

for r<a

$$= {}_2F_1\left\{\frac{1}{2}, -\frac{1}{2}; 1; \frac{r^2}{b^2}\right\} - \frac{a^2}{2b^2} \cdot \frac{b}{r} {}_2F_1\left\{\frac{1}{2}, \frac{1}{2}; 2; \frac{b^2}{r^2} \cdot \frac{a^2}{b^2}\right\}$$

for a<r<b

$$= \frac{b}{2r} {}_2F_1\left\{\frac{1}{2}, \frac{1}{2}; 2; \frac{b^2}{r^2}\right\} - \frac{a^2}{2b^2} \cdot \frac{b}{r} {}_2F_1\left\{\frac{1}{2}, \frac{1}{2}; 2; \frac{b^2}{r^2} \cdot \frac{a^2}{b^2}\right\}$$

for r>b   [34]

In the same manner, we can derive the average concentration over the surface of the ring

$$C_{Av} = C^\infty + \frac{2Qa}{D(b^2-a^2)} \int_0^\infty \int_a^b J_0(\lambda r) J_1(\lambda a) \, r dr \, \frac{d\lambda}{\lambda}$$

$$- \frac{2Qb}{D(b^2-a^2)} \int_0^\infty \int_a^b J_0(\lambda r) J_1(\lambda b) r dr \, \frac{d\lambda}{\lambda}$$

$$= C^\infty + \frac{2Qa}{D(b^2-a^2)} \int_0^\infty J_1(\lambda a) \left\{ bJ_1(\lambda b) - aJ_1(\lambda a) \right\} \frac{d\lambda}{\lambda^2}$$

$$- \frac{2Qb}{D(b^2-a^2)} \int_0^\infty J_1(\lambda b) \left\{ bJ_1(\lambda b) - aJ_1(\lambda a) \right\} \frac{d\lambda}{\lambda^2}$$

$$= C^\infty - \frac{2Qb^2}{D(b^2-a^2)} \int_0^\infty J_1^2(\lambda b) \frac{d\lambda}{\lambda^2} - \frac{2Qa^2}{D(b^2-a^2)} \int_0^\infty J_1^2(\lambda a) \frac{d\lambda}{\lambda^2}$$

$$+ \frac{4Qab}{D(b^2-a^2)} \int_0^\infty J_1(\lambda a) J_1(\lambda b) \frac{d\lambda}{\lambda^2}$$

$$= C^\infty - \frac{8Qb^3}{3\pi D(b^2-a^2)} - \frac{8Qa^3}{3\pi D(b^2-a^2)}$$

$$+ \frac{2Qa^2 b}{D(b^2-a^2)} {}_2F_1\left\{ \frac{1}{2}, -\frac{1}{2}; 2; \frac{a^2}{b^2} \right\} \qquad [35]$$

We note that this expression is exact. It holds for all values of b/a including that for a disk (b/a = ∞), from which we recover [29]

$$C_{Av} = C^\infty - \frac{8Qb}{3\pi D} \qquad [29]$$

Mass transfer coefficients to the surface of the ring can be derived in various ways depending on the position at which the surface concentration is defined. A convenient choice is the average concentration [35] and setting this to zero we obtain

$$(km)_Q = \frac{Q}{c^\infty} = \frac{D(b^2-a^2)}{\frac{8b^3}{3\pi} - \frac{8a^3}{3\pi} - 2a^2 b \, _2F_1\left\{\frac{1}{2},-\frac{1}{2};\, 2;\, \frac{a^2}{b^2}\right\}} \qquad [36]$$

The expression of the results in terms of hypergeometric functions is convenient as these series generally converge rapidly and their propagation is simple and well suited to representation by efficient algorithms.

Polarization curves
Disk electrodes

The interpretation of electrochemical data, such as of polarization curves raises the question whether it is appropriate to use the mass transfer coefficient for constant concentration or constant flux conditions (equations [23] and [30] respectively). We examine this question for the steady state polarization curve for the simplest example of a reaction following Butler-Volmer kinetics:

$$Q = \frac{i_0}{F} \left\{ \left[\frac{c^\infty - c}{c^\infty}\right] \exp\left[\frac{-\alpha \eta F}{RT}\right] - \left[\frac{c^\infty + c}{c^\infty}\right] \exp\left[\frac{(1-\alpha)\eta F}{RT}\right] \right\} \qquad [37]$$

and for equal concentrations of the oxidized and reduced species. The use of averaged concentrations at the surface allows us to use [37] as a boundary condition instead of [24]. We obtain

$$\frac{FQ}{i_o} + \frac{2Q}{DC^\infty} \int_0^\infty \left[J_1(\lambda a)\right]^2 \frac{d\alpha}{\alpha^2} \left\{\exp\left[\frac{-\alpha\eta F}{RT}\right] + \exp\left[\frac{(1-\alpha)\eta F}{RT}\right]\right\}$$

$$= \left\{\exp\left[\frac{-\alpha\eta F}{RT}\right] - \exp\left[\frac{(1-\alpha)\eta F}{RT}\right]\right\} \quad [38]$$

The use of [29] and [30] allows us to write this in this form

$$i = F(km)_Q C^\infty \frac{\left\{\exp\left[\frac{-\alpha\eta F}{RT}\right] - \exp\left[\frac{(1-\alpha)\eta F}{RT}\right]\right\}}{\left\{\frac{F(km)_Q C^\infty}{i_o} + \exp\left[\frac{-\alpha\eta F}{RT}\right] + \exp\left[\frac{(1-\alpha)\eta F}{RT}\right]\right\}} \quad [39]$$

The use of $(km)_Q$ is appropriate close to the reversible potential where the irreversibility of the reaction will ensure essentially uniform conditions over the surface (cf. tertiary current distribution) and where the concentration changes are small. However, [39] will predict the wrong limiting current since at that limit it is appropriate to assume that the concentration is uniform over the surface i.e., we should use $(km)_C$ at high η. Although $(km)_C$ can, strictly speaking, only be used for these limiting conditions, we can obtain an estimate of the magnitude of the likely error due to using $(km)_Q$ by expressing [39] as a fraction of the limiting current density

$$i_\ell = F(km)_C C^\infty \quad [40]$$

giving

$$\frac{i}{F(k_m)_c C^\infty} = \frac{(k_m)_Q}{(k_m)_c} \frac{\left\{\exp\left[\frac{-\alpha\eta F}{RT}\right] - \exp\left[\frac{(1-\alpha)\eta F}{RT}\right]\right\}}{\left\{\frac{F(k_m)_Q C^\infty}{i_0} + \exp\left[\frac{-\alpha\eta F}{RT}\right] + \exp\left[\frac{(1-\alpha)\eta F}{RT}\right]\right\}} \quad [41]$$

Figure 1A illustrates [41] for $\alpha = 0.5$ and for various values of $F(k_m)_Q C^\infty/i_0$. Evidently, if this parameter is sufficiently large we obtain kinetic control at low η and the assumption of an incorrect mass transfer coefficient at the limiting current will have no effect on the shapes of the kinetic plots. The effects of such an assumption are small even for essentially reversible processes, Figure 1B, and the maximum error for any part of the polarization curves is of order

$$\frac{(k_m)_c - (k_m)_Q}{(k_m)_c} \approx 7.5\% \quad [42]$$

Clearly, this error can be reduced to $\approx 3.75\%$ by using instead a mean transfer coefficient

$$k_m = \frac{(k_m)_c + (k_m)_Q}{2} \quad [43]$$

Ring electrodes

Figure 2 illustrates the concentration changes (expressed by the dimensionless variable $C_{z=0}$) over the range $0 < r < 3b$ at $z = 0$ for a disk ($b/a = \infty$) and for rings having $b/a = 2.0$ and 1.2 while Figure 3 illustrates the concentration changes

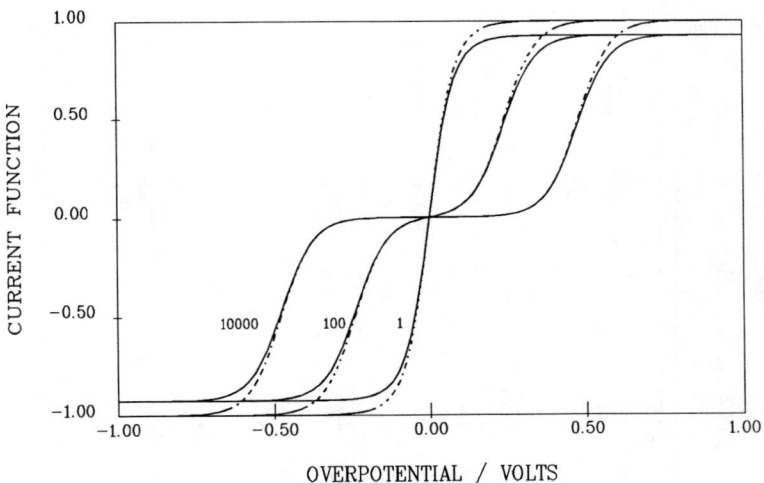

Figure 1A. Plot of the current parameter = LHS Equation [41] vs. overpotential for several values of the rate parameter $F(km)_Q c^\infty / i_0$. The plot for the constant concentration approximation is also included (---).

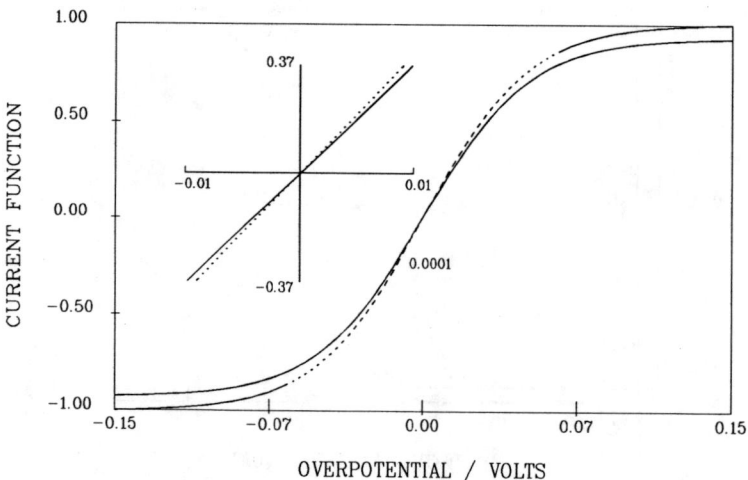

Figure 1B. Same as Figure 1A for a fast electron transfer reaction, (---) plot for $(k_m)_c$. (———) plot for Equation [41]. Rate parameter = 0.0001. Inset: expansion of the plot near the reversible potential.

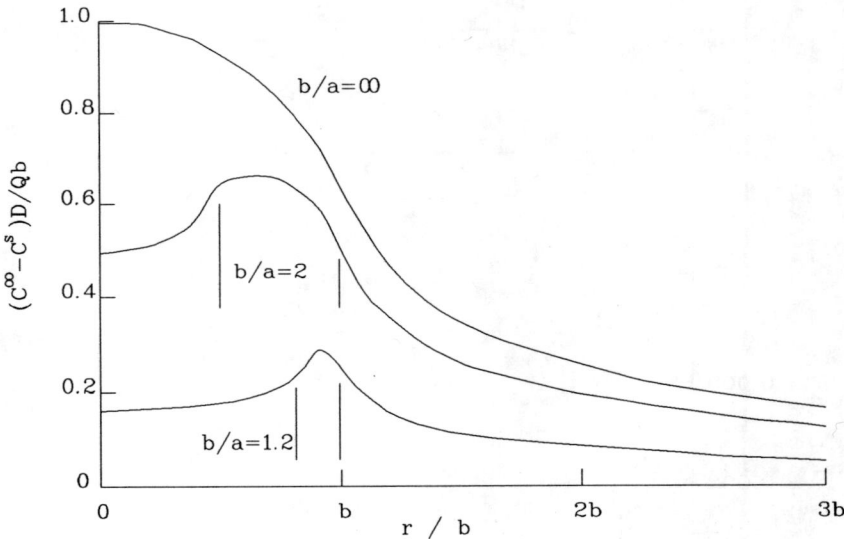

Figure 2. Plot of the dimensionless concentration changes $(c^\infty - c^s)D/Qb$ as a function of radial position for a disk ($b/a = \infty$) and for rings having the indicated radius ratios. The vertical bars indicate the dimensions for the rings.

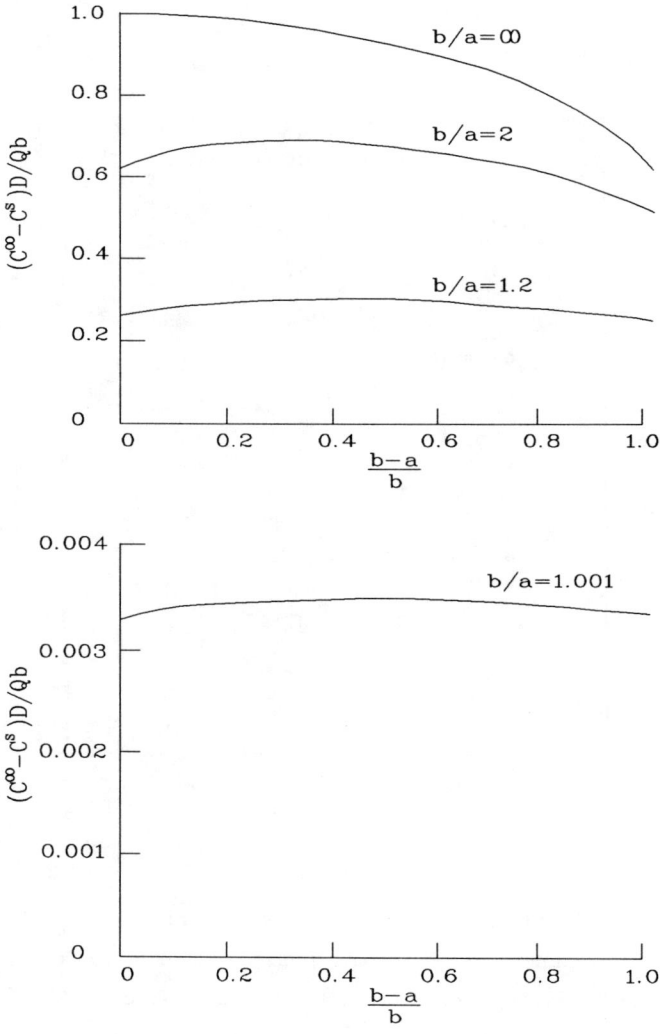

Figure 3. Plots of the dimensionless concentration changes $(C^\infty - C^s)D/Qb$ as a function of radial position over the surface or rings with the indicated radius ratio (b/a).

over the ring alone. These figures show that the concentration changes are markedly reduced as the thickness of the rings is decreased. At this time it is certainly feasible to prepare rings 10nm thick on fibers of 10µm radius giving $b/a \approx 1.001$. Figure 3 shows the consequent large increase of mass transfer to the ring which for rings of these dimensions reaches $(k_m)_Q \approx 2.5$ cm s^{-1}; for rings of the same thickness prepared on 1µm radius fibers $(k_m)_Q \approx 35$ cm s^{-1}. It is therefore doubtful whether there would be any benefit in carrying out transient measurements with microelectrodes of such dimensions.

The definition of mass transfer coefficients in terms of the average concentrations, equations [36] and [29], however, implies that some parts of the rings (or disks) will be at negative concentrations at the limiting current which is physically unsound. In this region it is best to base the definition on the attainment of $c^s_{z=0} = 0$ over the whole surface. Values can be derived from an exact calculation of the capacitance of a ring based on the use of triple integral equations (33) and these values differ appreciably from those for constant flux conditions, Table 1 columns 2 and 4. Table 1 also gives values calculated from an approximation given for constant flux conditions (34), column 3

$$\left[\frac{(k_m)_{ring}}{(k_m)_{disk}}\right]_Q = \frac{8}{3\left(1-\frac{a}{b}\right)\ln\left\{4e^{3/2}\left(\frac{1+a/b}{1-a/b}\right)\right\}} \qquad [44]$$

as well as values calculated for an approximation for constant concentration conditions (35), column 5.

Table 1

The ratios

$$\left[\frac{(km)_{ring}}{(km)_{disk}}\right]_Q \qquad \left[\frac{(km)_{ring}}{(km)_{disk}}\right]_C \qquad \left[\frac{(km)_{ring,Q}}{(km)_{ring,C}}\right]$$

as a function of the radius ratio b/a.

b/a	$\left[\dfrac{(km)_{ring}}{(km)_{disk}}\right]_Q$		$\left[\dfrac{(km)_{ring}}{(km)_{disk}}\right]_C$		$\left[\dfrac{(km)_{ring,Q}}{(km)_{ring,C}}\right]$
	using [36] and [29]	using [44]	using data ref (33)	using [45]	
∞	1.000	not valid			.9253 (= $\frac{3\pi^2}{32}$)
2	1.3514	1.3039	1.3080	1.2747	.9560
1.5	1.7859	1.7795	1.7089	1.6892	.9670
1.25	2.6259	2.6229	2.4933	2.4824	.9745
1.2	3.0303	3.0279	2.8721	2.8633	.9762
1.125	4.1976	4.1962	3.9671	3.9614	.9790
1.091	5.3110	5.3100	5.0112	5.0078	.9806
1.021	17.396	17.347	16.376	16.345	.9829
1.010	32.887	32.887		30.858	.9861
1.005	60.358	60.358		56.573	.9872
1.001	254.52	254.52		238.08	.9892

$$\left[\frac{(k_m)_{ring}}{(k_m)_{disk}}\right]_c = \frac{\pi^2}{4\left(1-\frac{a}{b}\right)\ln\left\{16\left(\frac{1+a/b}{1-a/b}\right)\right\}} \quad [45]$$

The differences between the values for the exact analyses of the problem using constant concentration and constant flux conditions, columns 2 and 4, illustrate the difficulties of choosing an appropriate model (compare previous section). However, Figure 3 shows that decreasing ring thickness leads to progressively more uniform concentrations over the surface of the ring which for $b/a = 1.001$ shows a maximum change of $\approx 5\%$ between the edges and center $(k_m)_Q$ therefore approaches $(k_m)_c$ as $b/a \to 1$, Table 1 column 6. It is clear from the data in Figures 2 and 3 as well as the data in Table 1 that the best strategy for many types of experiments at this stage of development of the subject is to use microring electrodes having the smallest attainable thickness since this will minimize the differences between the approximations. This strategy also maximizes the mass transfer coefficients in comparison with those achievable at disk electrodes, Table 1 columns 2 and 4.

Approximate treatments of experimental data

It will be shown in the next section that it is possible to extend the analyses outlined in the previous sections and in the literature to the discussion of many types of experiments provided the reaction kinetic terms are first order or pseudo-first order. It is usually necessary to make approximations in the modelling although the mathematical derivations are exact. For other types of experiments it is frequently necessary to use some approximate methods of analyzing the data. The answers derived may, of course, eventually be shown to be correct.

We illustrate the application of such techniques with the simple example of a c.e. reaction (36, 37)

$$A \underset{k_2}{\overset{k_1}{\rightleftarrows}} B \qquad (i)$$

$$B \xrightarrow{ne} Product \qquad (ii)$$

It should be noted that, as it is easy to maintain high rates of diffusion to the surfaces of microelectrodes, the forward reaction will be effectively pseudo-zero order so that we will usually only need to solve the single equation

$$\frac{\partial C}{\partial t} = D \frac{\partial^2 C}{\partial r^2} + \frac{D}{r} \frac{\partial C}{\partial r} + D \frac{\partial^2 C}{\partial z^2} + k_1 - k_2 c \qquad [46]$$

$$= 0 \text{ in the steady state}$$

Instead of solving [46], we solve the equivalent problem for a microspherical electrode i.e.,

$$D \frac{\partial^2 C}{\partial r^2} + \frac{2D}{r} \frac{\partial C}{\partial r} + k_1 - k_2 C = 0 \qquad [47]$$

with $C = C^\infty = \dfrac{k_1}{k_2}$, $r = \infty$ [48]

and the simplest condition

$C = 0$, $r = 0$ [49]

We obtain

$$C = \frac{k_1}{k_2}\left\{1 - \frac{a}{r}\exp\left[-\left(\frac{k_2}{D}\right)^{1/2}(r-a)\right]\right\}$$ [50]

and a flux (per unit area) of the microelectrode

$$F = D\left[\frac{dC}{dr}\right]_{r=a} = \frac{Dk_1}{k_2}\left\{\frac{1}{a} + \left(\frac{k_2}{D}\right)^{1/2}\right\}$$ [51]

We next correct the flux to the sphere to that at a disk by assuming that the same relationship holds as for the case of a reversible reaction viz

$$a_{disk} = \frac{4}{\pi}a_{sphere}$$ [52]

This gives

$$F = D\left[\frac{dC}{dz}\right]_{z=0} = \frac{4Dk_1}{\pi a_{disk}}\left\{1 + \frac{\pi}{4}\left[\frac{k_2 a_{disk}^2}{D}\right]^{1/2}\right\}$$ [53]

Problems of this type always give rise to simple "working curves" which in this case are fitted by determining the flux as a function of the radius of the microelectrode.

Extension of the approach to the discussion of the non-steady state behavior: chronopotentiometry

It has been noted above that experiments with very small electrodes (dimensions $\approx 0.5\mu$) will usually be carried out in the steady state (or under quasi-steady state conditions). The measurement of transient responses with readily available instrumentation becomes feasible for electrodes having somewhat larger dimensions. The approach outlined above can easily be adopted to predict the transient behavior and to devise a number of novel experiments.

For the simplest types of experiments involving a single reactant we need to solve

$$\frac{\partial C}{\partial t} = D \frac{\partial^2 C}{\partial r^2} + \frac{D}{r} \frac{\partial C}{\partial r} + D \frac{\partial^2 C}{\partial z^2} \qquad [54]$$

subject to the initial condition

$$C = C^\infty, \quad r > 0, \quad z > 0, \quad t = 0 \qquad [55]$$

Laplace transformation gives

$$\frac{\partial^2 \overline{C}}{\partial r^2} + \frac{1}{r} \frac{\partial \overline{C}}{\partial r} + \frac{\partial^2 \overline{C}}{\partial z^2} - q^2 \overline{C} + \frac{C^\infty}{D} = 0 \qquad [56]$$

where $q^2 = \dfrac{s}{D}$ [57]

and s is the variable of the transformation. We therefore seek the solution for the complementary function from

$$\dfrac{\partial^2 \overline{C}_{C.F.}}{\partial r^2} + \dfrac{1}{r}\dfrac{\partial \overline{C}_{C.F.}}{\partial r} + \dfrac{\partial^2 \overline{C}_{C.F.}}{\partial z^2} - q^2 \overline{C}_{C.F.} = 0 \qquad [58]$$

It is possible to suggest a number of substitutions

$$\overline{C}_{C.F.} = \overline{v}\, \exp[-f(\lambda q)z] \qquad [59]$$

each of which converts [58] to

$$\dfrac{d^2 \overline{v}}{dr^2} + \dfrac{1}{r}\dfrac{d\overline{v}}{dr} + \alpha^2 \overline{v} = 0 \qquad [60]$$

where $\alpha^2 = [f(\lambda,q)]^2 - q^2$ [61]

For the simple case of chronopotentiometry and for a constant flux of strength $-Q$ over the surface of the disk, we therefore obtain a result which is formally similar to [28] namely

$$\overline{C} = \dfrac{C^\infty}{s} - \dfrac{Qa}{Ds}\int_0^\infty \exp[-f(\lambda,q)z] J_0(\alpha r) J_1(\alpha a)\dfrac{d\alpha}{f(\lambda,q)} \qquad [62]$$

and, at $z = 0$,

$$\bar{C} = \frac{C^\infty}{s} - \frac{Qa}{Ds}\int_0^\infty J_0(\alpha r)J_1(\alpha a)\frac{d\alpha}{f(\lambda,q)} \qquad [63]$$

We observe that the interpretation of the arguments of the Bessel functions depends on the nature of the assumption of the form of $f(\lambda,q)$ and, indeed, on the nature of the experiment (e.g., see chronoamperometry below). The use of the simple form

$$f(\lambda,a) = (\alpha^2 + q^2)^{1/2} \qquad [64]$$

gives

$$\bar{C} = \frac{C^\infty}{s} - \frac{Qa}{Ds}\int_0^\infty J_0(\alpha r)J_1(\alpha a)\frac{d\alpha}{(\alpha^2+q^2)^{1/2}} \qquad [65]$$

for galvanostatic conditions where α is now independent of q. Consequently we can invert immediately to the t-domain:

$$C = C^\infty - \frac{Qa}{D}\int_0^\infty J_0(\alpha r)J_1(\alpha a)\,\mathrm{erf}(D^{1/2}\alpha t^{1/2})\frac{d\alpha}{\alpha} \qquad [66]$$

where erf(y) denotes the error function. We can evaluate the average concentration at $z = 0$

$$C_{Av} = C^\infty - \frac{2Q}{D}\int_0^\infty [J_1(\alpha a)]^2\,\mathrm{erf}(D^{1/2}\alpha t^{1/2})\frac{d\alpha}{\alpha^2} \qquad [67]$$

ERRATA.

The following equations are misprinted in Chapter 2.

$$z(z)\frac{d^3r}{dz^3} + \frac{2(z)}{r}\frac{dr}{dz} + R(z)\frac{d^2z(z)}{dz^2} = 0 \qquad [3]$$

$$\int_0^\infty \sin(\lambda a)J_o(\lambda r)d\lambda = \left\{\begin{array}{ll}\frac{1}{(a^2-r^2)^{1/2}}, & r < a \\ 0, & r > a\end{array}\right. \qquad \begin{array}{c}[16]\\ [17]\end{array}$$

$$C_{av} - C^\infty = -\frac{D\alpha}{2Q}\int_0^\infty\int_0^a J_o(\lambda r)J_1(\lambda a)\frac{r}{a}\,dr\,d\lambda - C^\infty - \frac{2Q}{Da}\int_0^\infty[J_1(\lambda a)]^2\,d\lambda\frac{2}{A} \qquad [29]$$

$$\frac{PQ}{2Q} + \frac{1}{A}\int_0^\infty[J_1(\lambda a)]^2\frac{d\phi}{a^2}\left\{\exp\left[\frac{-Vq}{RT}\right]dxe + \exp\left[\frac{-(1-\alpha)Vq}{RT}\right]dxe\right\}$$
$$- \left\{\exp\left[\frac{-Vq}{RT}\right]dxe - \exp\left[\frac{-(1-\alpha)Vq}{RT}\right]dxe\right\} \qquad [38]$$

$$f(\lambda,q) = (\alpha^2 + q^2)^{1/2} \qquad [64]$$

$$Q(s) = -\frac{\left[C^\infty-C_{Av}\right]a\frac{D^{1/2}}{as^{1/2}}}{\int_0^\infty\left[\left[J_1\left(\frac{\alpha as^{1/2}}{D^{1/2}}\right)\right]^2\right]\frac{d\phi}{\phi(\phi^2+1)^{1/2}}} \qquad [74]$$

Table 3 actually lists $\left[D^{1/2}/\phi_2 as^{1/2}\right]$ (or $\left[D^{1/2}/\phi_2 ak_2\right]$). Multiply the inverse of the tabulated values by $\left[D^{1/2}/as^{1/2}\right]$ (or $\left[D^{1/2}/ak_2\right]$) to obtain ϕ_2.

(Please see reverse)

Table 2 is expanded and corrected to:

Dε/a²	φ₁(Dε/a²)
1.0000e-6	2.1217e-4
1.7778e-6	2.5460e-4
2.3464e-6	2.9702e-4
3.1623e-6	3.9444e-4
5.6250e-6	3.8186e-4
9.0000e-6	4.2427e-4
1.1111e-5	8.4826e-4
2.5000e-5	1.2720e-3
2.5000e+4	1.6954e-3
4.6964e-4	2.1185e-3
5.6250e-4	2.5414e-3
3.8083e-3	2.9640e-3
4.2300e-3	3.3863e-3
8.4318e-3	3.8083e-3
1.2605e-2	4.2300e-3
1.6751e-2	1.1111e+3
2.0868e-2	5.0000e+3
2.4957e-2	1.7778e+3
2.9018e-2	2.5000e+2
3.3051e-2	5.6250e+2
3.7055e-2	2.3464e+2
4.1032e-2	1.0000e+2
7.9259e-2	5.0000e+1
1.1472e-1	1.1111e+1
1.4749e-1	2.5000e+0
1.7766e-1	1.0000e+0
2.0537e-1	1.7778e+0
2.3064e-1	4.0400e+0
2.5372e-1	5.6250e+0
2.7473e-1	2.3464e+0
2.9381e-1	1.0000e+0
4.1045e-1	5.0000e-1
4.1965e-1	1.1111e-1
4.8524e-1	2.5000e-1
5.0085e-1	1.0000e-1
5.1132e-1	1.7778e-1
5.1883e-1	4.0400e-2
5.2448e-1	5.6250e-2
5.2888e-1	2.3464e-2
5.3240e-1	1.0000e-2
5.4828e-1	5.0000e-3
5.5385e-1	1.1111e-3
5.5623e-1	2.5000e-4
5.5783e-1	1.0000e-4
5.5889e-1	1.7778e-4
5.5964e-1	4.0400e-5
5.6021e-1	5.6250e-5
5.6065e-1	2.3464e-5
5.6101e-1	1.0000e-5
5.6260e-1	2.5000e-5
5.6134e-1	1.1111e-5
5.6340e-1	2.5000e-6
5.6356e-1	1.0000e-6

Table 4 is expanded and corrected to:

Dε/a²	φ₃(Dε/a²)
2.0000e-2	7.5112e-5
3.1623e-2	1.8778e-4
3.5355e-2	2.6826e-4
3.7796e-2	3.7556e-4
4.4721e-2	4.6945e-4
5.0000e-2	9.3831e-4
7.0711e-2	1.8728e-3
1.0000e-1	2.3085e-3
1.1180e-1	2.6706e-3
1.1952e-1	3.7299e-3
1.4142e-1	4.6526e-3
1.5811e-1	9.2073e-3
2.3611e-1	1.8023e-2
3.1623e-1	2.2285e-2
3.5355e-1	2.5269e-2
3.7796e-1	3.4486e-2
4.4721e-1	4.2138e-2
5.0000e-1	7.4851e-2
7.0711e-1	1.1751e-1
1.0000e+0	1.3151e-1
1.1180e+0	1.3971e-1
1.1952e+0	1.5820e-1
1.4142e+0	1.7040e-1
1.5811e+0	2.2654e-1
1.9521e+0	2.6954e-1
2.3611e+0	2.9696e-1
3.1623e+0	3.1727e-1
3.5355e+0	3.5219e-1
3.7796e+0	3.6511e-1
4.4721e+0	3.9321e-1
5.0000e+0	4.0842e-1
7.0711e+0	4.1156e-1
1.0000e+1	4.1313e-1
1.1180e+1	4.1631e-1
1.1952e+1	4.1792e-1
1.4142e+1	4.2116e-1
1.5811e+1	4.2277e-1
1.9521e+1	4.2327e-1
2.3611e+1	4.2360e-1
3.1623e+1	4.2378e-1
3.5355e+1	4.2406e-1
3.7796e+1	4.2440e-1
4.4721e+1	4.2440e-1
5.0000e+2	4.2418e-1
7.0711e+2	4.2418e-1
1.0000e+2	4.2418e-1
1.1180e+2	4.2418e-1
1.1952e+2	4.2436e-1
1.4142e+2	4.2418e-1
3.1623e+3	4.2416e-1
1.0000e+4	4.2441e-1

With the substitutions

$$\ell^2 = Dt \qquad [68]$$

$$\beta = \alpha \ell \qquad [69]$$

[67] can be written in terms of dimensionless variables and parameters

$$C_{Av} = C^\infty - \frac{2Qa}{D} \cdot \frac{\ell}{a} \int_0^\infty \left[J_1\left(\frac{\beta a}{\ell}\right) \right]^2 \mathrm{erf}(\beta) \frac{d\beta}{\beta^2}$$

$$= C^\infty - \frac{2Qa}{D} \cdot \Phi_1\left(\frac{Dt}{a^2}\right) \qquad [70]$$

The function Φ_1 is tabulated in Table 2 as a function of the dimensionless parameter (Dt/a^2).

At sufficiently long times and provided Q is sufficiently small we recover [29]. If Q is sufficiently large we will get a sharp transition time determined by

$$\frac{2Qa}{DC^\infty} \cdot \frac{\ell}{a} \int_0^\infty \left[J_1\left(\frac{\beta a}{\ell}\right) \right]^2 \mathrm{erf}(\beta) \frac{d\beta}{\beta^2} = 1 \qquad [71]$$

Figure 4 illustrates the square root of the dimensionless transition times $D\tau/a^2$ as a function of the inverse of the dimensionless flux $2Qa/DC^\infty$. The values are close to those predicted by Aoki and Osteryoung (17) even though these authors based their analysis on the uniform surface concentration boundary condition, [11]; the expressions derived could not be inverted exactly over the entire time range whereas [71] is exact. It should be noted that the constant flux condition is

Table 2. Values of the function $\Phi_1(Dt/a^2)$, equation [70].

Dt/a^2	$\Phi_1(Dt/a^2)$	Dt/a^2	$\Phi_1(Dt/a^2)$
1000.0	0.81387	0.05000	0.02087
100.00	0.56351	0.04762	0.01989
50.00	0.55845	0.04545	0.01900
33.33	0.55492	0.04348	0.01819
25.00	0.55162	0.04167	0.01744
20.00	0.54838	0.04000	0.01675
16.667	0.54517	0.03846	0.01611
14.286	0.54197	0.03704	0.01553
10.000	0.53242	0.03571	0.01498
5.0000	0.50086	0.03448	0.01447
3.3333	0.46980	0.03333	0.01399
1.0000	0.29382	0.03226	0.01354
0.5000	0.17766	0.03125	0.01313
0.3333	0.12594	0.03030	0.01273
0.2500	0.09733	0.02857	0.01201
0.2000	0.07926	0.02778	0.01168
0.16667	0.06683	0.02703	0.01137
0.14286	0.05776	0.02632	0.01107
0.12500	0.05085	0.02564	0.01079
0.11111	0.04542	0.02500	0.01052
0.10000	0.04103	0.02439	0.01027
0.09091	0.03742	0.02381	0.01003
0.08333	0.03439	0.0233	0.00979
0.07692	0.03181	0.0227	0.00957
0.07143	0.02960	0.0222	0.00936
0.06667	0.02767	0.0217	0.00916
0.06250	0.02598	0.0213	0.00897
0.05882	0.02448	0.0208	0.00878
0.05556	0.02314	0.0204	0.00860
0.05263	0.02195	0.0200	0.00843

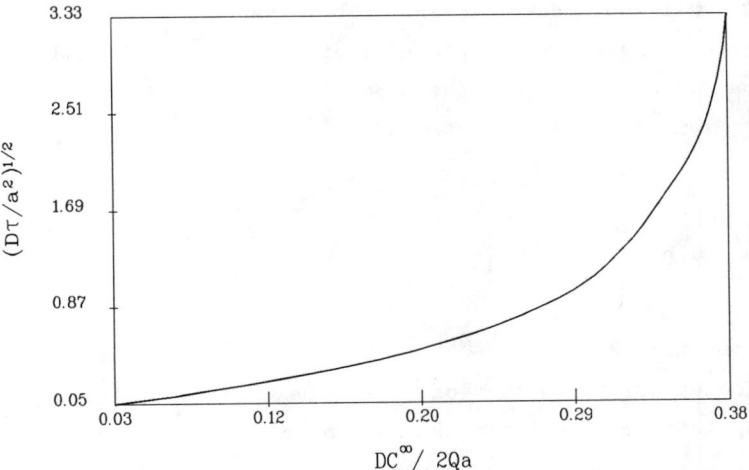

Figure 4. Plot of the square root of the dimensionless transition time $(D\tau/a^2)^{1/2}$ as a function of the inverse of the dimensionless flux $(2Qa/DC^{\infty})$.

more likely to apply for most of the duration of the experiment rather than the constant surface concentration condition; the agreement between the transition times derived using the two approaches shows that the interpretation is not very sensitive to the nature of the assumptions (compare section on the interpretation of polarization curves). Figure 4 shows the expected linear dependence at high values of $2Qa/DC^\infty$ where we observe essentially linear diffusion to the electrode followed by a rapid rise at low values of $2Qa/DC^\infty$ as this parameter approaches the condition required for the observation of a steady state

$$\frac{2Qa}{DC^\infty} \leq \frac{4}{3\pi} = 0.4244 \qquad [72]$$

(cf. equation [29]).

The calculated values of C_{Av} may be used to derive the chronopotentiometric responses for chosen models of the electrode reactions. Thus assuming simple Butler-Volmer kinetics, (37), we obtain

$$\frac{FQ}{i_o} + \frac{2Qa}{DC^\infty} \Phi_1\left[\frac{Dt}{a^2}\right] \left\{\exp\left[\frac{-\alpha\eta F}{RT}\right] + \exp\left[\frac{(1-\alpha)\eta F}{RT}\right]\right\}$$

$$= \exp\left[\frac{-\alpha\eta F}{RT}\right] - \exp\left[\frac{(1-\alpha)\eta F}{RT}\right] \qquad [73]$$

which shows that the transients are a function of α, FQ/i_o and $2Qa/DC^\infty$.

Chronoamperometry

Previous discussions of this case have shown that the use of the boundary conditions [11] and [12] causes considerable difficulties. In the context of the analyses developed here we can trace these to the "mixed" nature of these boundary conditions, the concentration being specified over $0 < r < a$ and the flux for $r > a$. Thus the application of [14] and [15] to the nonsteady state destroys the discontinuity [16] and [17] (although the behavior is correct at the short and long time limits). We can, however, make substantial changes to the integral [28] without affecting [25]-[27]. Thus we can use [65] and pose the question: what strength of sink, $Q(s)$ in the s-domain will give a constant average surface concentration over the sink? Evidently this is determined from

$$Q(s) = \frac{\dfrac{\left(C^{\infty} - C_{Av}\right) a}{2} \dfrac{D^{1/2}}{as^{1/2}}}{\displaystyle\int_0^{\infty} \left[J_1\left(\beta \dfrac{as^{1/2}}{D^{1/2}}\right)\right]^{1/2} \dfrac{d\beta}{\beta \left(\beta^2 + 1\right)^{1/2}}} \qquad [74]$$

$$= \frac{\left(C^{\infty} - C_{Av}\right) a}{2} \Phi_2 \left(\frac{as^{1/2}}{D^{1/2}}\right) \qquad [75]$$

and where β is now defined by

$$\beta = \frac{D^{1/2} \alpha}{s^{1/2}} \qquad [76]$$

Table 3 gives Φ_2 as a function of $(as^{1/2}/D^{1/2})$. It should be noted that the argument of the Bessel function is now dependent on s. Inversion at short s (long t) gives

$$Q = \frac{3\pi D}{8a} (C^\infty - C_{Av}) \tag{77}$$

(cf. [30]) while at long s (short t) we obtain the Cottrell behavior

$$Q = \frac{D^{1/2}(C^\infty - C_{Av})}{\pi^{1/2} t^{1/2}} \tag{78}$$

(since we have planar diffusion to the disk under these conditions) Numerical inversion at intermediate times gives $Q(t)$; the values are close to those which have been calculated by other algebraic and numerical procedures (12-16,18).

Other applications
Reactions in solution coupled to electrode processes: the c.e. reaction

The approaches outlined above for the discussion of the chronoamperometric response can be readily adapted for the interpretation of other types of experiments and we illustrate this with the example of the simple c.e. reaction (i) and (ii). Laplace transformation of [46] gives

$$\frac{\partial^2 \overline{C}}{\partial r^2} + \frac{1}{r}\frac{\partial \overline{C}}{\partial r} + \frac{\partial^2 \overline{C}}{\partial z^2} - \left(q^2 + \frac{k_2}{D}\right)\overline{C} + \frac{k_1}{D}\left(\frac{1}{k_2} + \frac{1}{s}\right) = 0 \tag{79}$$

Table 3. Values of the function $\Phi_2(as^{1/2}/D^{1/2})$

(or $\Phi_2(ak_2^{1/2}/D^{1/2})$)

equations [74] and [75] (equations [85] and [86]).

$as^{1/2}/D^{1/2}$ or $ak_2^{1/2}/D^{1/2}$	$\Phi_2(as^{1/2}/D^{1/2})$ or $\Phi_2(ak_2^{1/2}/D^{1/2})$	$as^{1/2}/D^{1/2}$ or $ak_2^{1/2}/D^{1/2}$	$\Phi_2(as^{1/2}/D^{1/2})$ or $\Phi_2(ak_2^{1/2}/D^{1/2})$
0.01000	0.004218	0.7000	0.2052
0.02000	0.008389	0.8000	0.2239
0.03000	0.01251	0.9000	0.2408
0.04000	0.01658	1.000	0.2561
0.05000	0.02061	2.000	0.3524
0.06000	0.02459	3.000	0.3972
0.07000	0.02852	4.000	0.4217
0.08000	0.03241	5.000	0.4370
0.09000	0.03625	6.000	0.4473
0.1000	0.04005	7.000	0.4548
0.2000	0.07572	8.000	0.4604
0.3000	0.1076	9.000	0.4647
0.4000	0.1360	10.00	0.4682
0.5000	0.1616	11.00	0.4711
0.6000	0.1846		

In this case use of the substitution

$$\overline{C}_{C.F.} = \overline{v} \exp\left[-\left(\alpha^2 + q^2 + \frac{k_2}{D}\right)^{1/2} z\right] \qquad [80]$$

gives

$$\overline{C} = \frac{k_1}{k_2 s} - \frac{Qa}{Ds} \int_0^\infty J_o(\alpha r) J_1(\alpha a) \frac{d\alpha}{\left(\alpha^2 + q^2 + \frac{k_2}{D}\right)^{1/2}} \qquad [81]$$

for a constant flux Q at the surface. On setting the average concentration equal to zero we obtain

$$Q(s) = \frac{\dfrac{k_1 D^{1/2}}{2k_2 s}}{\displaystyle\int_0^\infty \left[J_1(\alpha a)\right]^2 \frac{d\alpha}{\alpha\left(D\alpha^2 + q^2 + k_2\right)^{1/2}}} \qquad [82]$$

If we redefine

$$\beta = \frac{D^{1/2} \alpha}{(s + k_2)^{1/2}} \qquad [83]$$

we can write [82] as

$$Q(s) = \frac{\frac{k_1 a}{2k_2} \cdot a\left(\frac{s+k_2}{D}\right)^{1/2}}{\frac{a^2 s}{D} \int_0^\infty \left[J_1\left(\beta a\left(\frac{s+k_2}{D}\right)^{1/2}\right)\right]^2 \frac{d\beta}{\beta(\beta^2+1)^{1/2}}} \qquad [84]$$

We therefore have a family of curves of $Q(s)$ as a function of $(a^2 s/D)$ each determined by a given value of the parameter $(a^2 k_2/D)$. Numerical inversion gives the responses in the t-domain. Here we restrict attention to the steady state limit

$$Q_{t\to\infty} = \frac{\frac{k_1 a}{2} \cdot \left(\frac{D}{a^2 k_2}\right)^{1/2}}{\int_0^\infty \left[J_1\left(\beta \frac{ak_2^{1/2}}{D^{1/2}}\right)\right]^2 \frac{d\beta}{\beta(\beta^2+1)^{1/2}}} \qquad [85]$$

which is identical to the chronoamperometric response [74] provided we replace $(c^\infty - C_{Av})$ by k_1 and $s^{1/2}$ by $k_2^{1/2}$ i.e., we obtain

$$Q = \frac{k_1 a}{2} \Phi_2\left(\frac{ak_2^{1/2}}{D^{1/2}}\right) \qquad [86]$$

The variation of Q with $(ak_2^{1/2}/D^{1/2})$ given by [86] is close to that calculated by the approximate approach leading to [51]. It should be noted, however, that the calculation outlined here is itself approximate in that we set the average concentration at the surface equal to zero.

Linear sweep amperometry

The application of galvanostatic conditions to microdisk electrodes is not particularly useful as the attainment of steady state diffusion complicates the observation of transition times. It is therefore more straightforward to apply a time dependent flux to the surface such as

$$Q = \gamma t \qquad [87]$$

The discussion of the chronopotentiometric response is readily modified to take account of [87]. For example, we derive the transition time

$$\frac{4\gamma t a}{DC^\infty} \cdot \frac{\ell}{a} \int_0^\infty \left[J_1\left(\beta\frac{a}{\ell}\right)\right]^2 \left[\int_0^\beta y \, \text{erf}(y) \, dy\right] \frac{d\beta}{\beta^4} = 1 \qquad [88]$$

or $\quad \dfrac{4\gamma t a}{DC^\infty} \cdot \Phi_3\left[\dfrac{Dt}{a^2}\right] = 1 \qquad [89]$

Table 4 gives values of Φ_3 as a function of Dt/a^2.

The A.C. impedance

In this case we can modify the boundary condition [24] for example to

$$D\frac{\partial C}{\partial z} = Q \sin \omega t, \qquad 0 < r < a, \quad z = 0, \quad t > 0 \qquad [90]$$

i.e.,

Table 4. Values of the function $\Phi_3(Dt/a^2)$, equations [88] and [89].

Dt/a^2	$\Phi_3(Dt/a^2)$	Dt/a^2	$\Phi_3(Dt/a^2)$
0.000	0.1881	156.2	0.01609
0.02041	0.1862	204.1	0.01416
0.02778	0.1817	277.8	0.01223
0.04000	0.1767	400.0	0.01025
0.06250	0.1708	625.0	0.008265
0.1111	0.1628	1111.	0.006239
0.2500	0.1493	2500.	0.004187
1.000	0.1161	10000.	0.002108
6.250	0.06552	15620	0.001268
11.11	0.05233	27780	0.001268
25.00	0.03720	62500	0.0008466
100.00	0.01798	100000	0.0002120
123.4	0.01798		

$$\frac{\partial \overline{C}}{\partial z} = \frac{Q\omega}{D(s^2 + \omega^2)}, \quad 0 < r < a, \quad z = 0 \qquad [91]$$

We obtain

$$\overline{C} = \frac{Qa}{D^{1/2}} \int_0^\infty J_o(\alpha r) J_1(\alpha a) \frac{\omega d\alpha}{(s^2 + \omega^2)(s + D\alpha^2)^{1/2}} \qquad [92]$$

The A.C. response is determined by the poles at

$$s = \pm j\omega \qquad [93]$$

and, writing

$$\ell^2 = \frac{D}{\omega} \qquad [94]$$

with again

$$\beta = \alpha \ell \qquad [69]$$

and taking the average concentration over $0 < r < a$ we obtain

$$C_{Av} = \frac{2Q}{D^{1/2} \omega^{1/2}} \int_0^\infty \left[J_1\left(\beta \frac{a}{\ell}\right) \right]^2 \frac{\sin(\omega t - \Theta/2) \, d\beta}{\beta(1 + \beta^4)^{1/4}} \qquad [95]$$

where

$$\tan\Theta = \frac{1}{\beta^2} \qquad [96]$$

Substitution of [95] in the linearized form of the Butler-Volmer equation [37] gives the electrode impedance in the usual way. Neglecting the double layer capacitance we obtain

$$Z' = \frac{RT}{\pi n F i_o a^2} + \frac{2RT}{\pi n^2 F^2 D^{1/2} \omega^{1/2} a^2 C^\infty} \int_0^\infty \left[J_1\left(\beta \frac{a}{\ell}\right) \right]^2 \frac{\cos(\Theta/2)\, d\beta}{\beta(1+\beta^4)^{1/4}} \quad [97]$$

$$Z'' = \frac{2RT}{\pi n^2 F^2 D^{1/2} \omega^{1/2} a^2 C^\infty} \int_0^\infty \left[J_1\left(\beta \frac{a}{\ell}\right) \right]^2 \frac{\sin(\Theta/2)\, d\beta}{\beta(1+\beta^4)^{1/4}} \quad [98]$$

Figure 5 gives a Cole-Cole plot of the dimensionless impedances

$$\frac{\pi n^2 F^2 D a\, C^\infty}{2RT} Z' = \frac{\pi n F D C^\infty}{2 a i_o} + 2 \left[\frac{D}{\omega a^2}\right]^{1/2} \Phi_4\left(\frac{a^2 \omega}{D}\right) \quad [99]$$

$$\frac{\pi n^2 F^2 D a\, C^\infty}{2RT} Z'' = 2 \left[\frac{D}{\omega a^2}\right]^{1/2} \Phi_5\left(\frac{a^2 \omega}{D}\right) \quad [100]$$

Φ_4 and Φ_5 being tabulated in Table 5 as a function of $(a^2\omega/D)$. As expected, the plot differs markedly from the plot for a large planar electrode (38). At low frequencies Z'' vanishes as the transport impedance becomes determined by the steady state mass transfer coefficient to the microelectrode surface. At sufficiently high frequencies the results resemble the familiar plot of a Warburg impedance and, as $\omega \to \infty$, $Z'' \to 0$ and Z' is determined by the charge transfer resistance. In contrast to measurements with conventional electrodes, the frequencies are scaled by the parameter (D/a^2). As in other experiments with microelectrodes the kinetics of fast reactions become

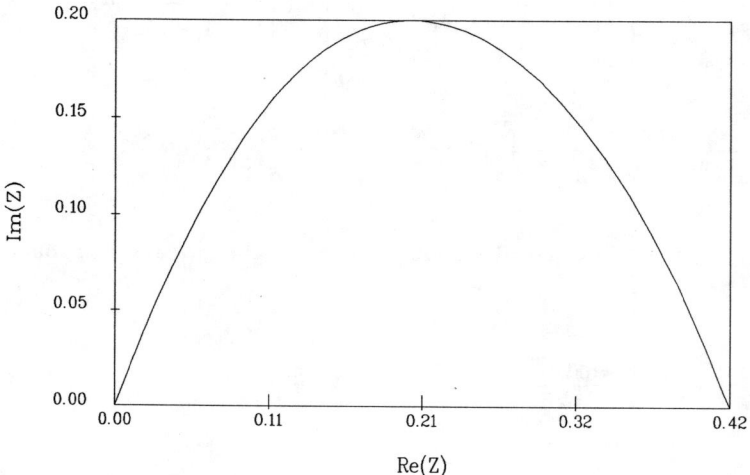

Figure 5. Cole-Cole plot of the dimensionless impedances, Equation [99] and [100].

Table 5. Values of the function $\Phi_4(a^2\omega/D)$ and $\Phi_5(a^2\omega/D)$, Equations [97] - [100]

$a^2\omega/D$	$\Phi_4(a^2\omega/D)$	$\Phi_5(a^2\omega/D)$
0.000400	0.00841	0.000070
0.001600	0.01669	0.000276
0.003600	0.02483	0.000612
0.006400	0.03282	0.001075
0.01000	0.04068	0.001658
0.04000	0.07786	0.006213
0.09000	0.1116	0.01309
0.1600	0.1421	0.02179
0.2500	0.1695	0.03186
0.4900	0.2156	0.05465
0.6400	0.2348	0.06677
0.8100	0.2517	0.07905
1.000	0.2665	0.09131
1.210	0.2794	0.1034
1.690	0.3002	0.1266
1.960	0.3086	0.1376
2.250	0.3157	0.1480
2.890	0.3269	0.1674
3.240	0.3312	0.1762
3.610	0.3349	0.1845
4.840	0.3427	0.2064
5.760	0.3460	0.2186
6.760	0.3482	0.2292
7.840	0.3496	0.2384
9.000	0.3506	0.2464
10.24	0.3513	0.2534
11.56	0.3517	0.2595
12.96	0.3520	0.2649
14.44	0.3522	0.2696
16.00	0.3524	0.2739
25.00	0.3529	0.2899
36.00	0.3532	0.3005
64.00	0.3534	0.3138
81.00	0.3534	0.3182
100.0	0.3535	0.3217
144.0	0.3535	0.3270
196.0	0.3535	0.3308
256.0	0.3535	0.3336
324.0	0.3535	0.3359
400.0	0.3535	0.3376
506.2	0.3535	0.3394
625.0	0.3535	0.3408
756.2	0.3535	0.3420

measurable by making (D/a) sufficiently large.

The equivalent circuit can be designated by Figure 6 where M denotes the diffusional impedance of the microelectrode. It is unlikely that the uncompensated solution resistance, R_u, will have to be taken into account as the spherical potential field in the solution minimizes R_u (just as the spherical concentration field maximizes km). The time constant $R_u C_{dl}$ will usually be short compared to the shortest accessible value of ω^{-1}: the response at high frequencies will in fact be similar to those of planar electrodes being determined by the product $R_{CT} C_{dl}$.

Conclusion

The simple approach outlined here can be used to analyze a wide variety of electrochemical experiments at a reasonable level of approximation. Where predictions have been made for both constant surface concentration, [11], and constant flux, [24], boundary conditions, the results are closely similar. This suggests that more detailed modelling taking into account the effects of the distribution of potential in the solution will not greatly change the predicted behavior which should lie between these limiting cases. All the results obtained here for disk electrodes have been formulated in terms of definite integrals; however, these integrals are readily evaluated using efficient algorithms (programs are available on application).

The approach can readily be extended to other types of experiments using disk electrodes. The interpretation of cyclic voltammetric and cyclic amperometric data as well as the behavior of other types of reactions in solution coupled to electrode reactions and of a variety of experiments carried out with thin ring electrodes will be discussed elsewhere (39). We

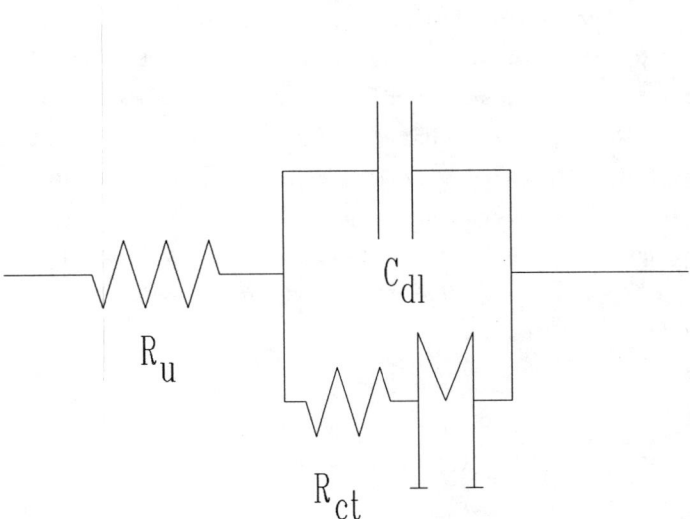

Figure 6. Equivalent circuit for a microelectrode.

conclude by commenting further on the use of the discontinuous integrals [25], [26] and [27] to predict the behavior for conditions of constant surface concentration. We have noted that in the case of time-dependent phenomena it is possible to predict the average surface concentration (e.g., [67]) whereas the direct application of [14] and [15] is not possible since we destroy the conditions for the discontinuity [16] and [17]. However, the question posed in deriving, for example, [74], can be extended. We can ask: what distribution of sources/sinks satisfying [25]-[27] is required to ensure a constant surface concentration? We illustrate one possible approach by considering the behavior in the steady state. A sink of strength $Q(a_1)$ constant over $0 < r < a_1$ generates a concentration distribution

$$\delta C(a_1, r) = \frac{-Q(a_1)a_1}{D} \int_0^\infty J_0(\alpha r) J_1(\alpha a_1) \frac{d\alpha}{\alpha} \qquad [101]$$

Superposition of all sinks in $0 < r < a$ gives a total concentration distribution

$$\int_0^a \delta C(a_1, r) da_1 = - \int_0^a \frac{Q(a_1)a_1}{D} \int_0^\infty J_0(\alpha r) J_1(\alpha a_1) \frac{d\alpha}{\alpha} da_1 \qquad [102]$$

and we require [102] to be independent of r and equal to $C^\infty - C^s$. We confine attention here instead to the special position $r = 0$ where

$$\int_0^\infty J_o(\alpha r) J_1(\alpha a_1) \frac{d\alpha}{\alpha} = 1 \qquad [103]$$

and pose the question: what distribution of sources do we require to make [102] independent of a? It is evident that the integral must reduce to a simple angle which would be derived from

$$Q(a_1) = \frac{2D(c^\infty - c^s)}{\pi a_1 (a^2 - a_1^2)^{1/2}} \qquad [104]$$

i.e., $\quad \dfrac{2(c^\infty - c^s)}{\pi} \displaystyle\int_0^a \dfrac{da_1}{(a^2 - a_1^2)^{1/2}} = c^\infty - c^s \qquad [105]$

If we equate the strength of the disk sink of radius a_1 to the local flux at this position, we recover [16] and integration gives the known result [20]. It is clear, therefore, that the starting equations [25]-[26] can be used to develop solutions for a variety of experiments modelled by conditions of constant concentration over the surface of the disk.

References

1. P. Bindra, A. P. Brown, M. Fleischmann and D. Pletcher, J. Electroanal. Chem., 58 (1975) 31.
2. P. Bindra, A. P. Brown, M. Fleischmann and D. Pletcher, J. Electroanal Chem., 58 (1975) 39.
3. G. Gunawardena, G. J. Hills and B. Scharifker, J. Electroanal Chem., 130 (1981) 99.
4. A. M. Bond, M. Fleischmann, S. B. Khoo, S. Pons and J. Robinson, Extended Abstracts 165th Meeting of the Electrochemical Society, May (1984), Electrochemical Society, Pennington N.J., (1984).
5. R. M. Wightman and K.R. Wehmeyer, Anal. Chem., 57 (1985) 1989.
6. M. Fleischmann, J. Ghoroghchian and S. Pons, J. Phys. Chem., 89 (1985) 5530.
7. M. Fleischmann, J. Ghoroghchian, S. Pons and D. Rolison, J. Phys. Chem., 90 (1986) 6392.
8. J. S. Symanski and S. Bruckenstein, Extended Abstracts 165th Meeting of the Electrochemical Society, May (1984), Electrochemical Society, Pennington, NJ, p. 527 (1984).
9. M. Fleischmann, S. Bandyopadhyay and S. Pons, J. Phys. Chem., 89 (1985) 5537.
10. J. B. Flanagan and L. Marcoux, J. Phys. Chem., 77 (1973) 1051.
11. M. Kakihana, H. Ikendi, G. P. Sato and K. Tokuda, J. Electroanal. Chem., 122 (1981) 19.
12. J. Heinze, J. Electroanal. Chem., 124 (1981) 73.
13. K.B. Oldham, J. Electroanal. Chem., 122 (1981) 1.
14. K. Aoki and J. Osteryoung, J. Electroanal. Chem., 122 (1981) 19.
15. B. Speiser and S. Pons, Can. J. Chem., 60 (1982) 1352.
16. B. Speiser and S. Pons, Can. J. Chem., 60 (1982) 2463.
17. K. Aoki, K. Akimoto, K. Tokuda, H. Matsuda and J. Osteryoung, J. Electroanal. Chem., 182 (1985) 281.
18. J. Cassidy and S. Pons, Can. J. Chem., 63 (1985) 3577.
19. T. Hepel, W. Plot and J. Osteryoung, J. Phys. Chem., 87 (1983) 1278.
20. T. Hepel and J. Osteryoung, J. Phys. Chem., 86 (1982) 1406.
21. D. Shoup and A. Szabo, J. Electroanal. Chem., 140 (1982) 237.
22. K. Aoki and J. Osteryoung, J. Electroanal. Chem., 160 (1984) 335.
23. M. Fleischmann and S. Pons, J. Electroanal. Chem., 222 (1987) 107.
24. M. Fleischmann and J. Harrison, Electrochim. Acta, 11 (1966) 749.
25. K. Aoki, K. Honda, K. Tokuda and H. Matsuda, J. Electroanal. Chem., 182 (1985) 267.

26. K. Aoki, K. Honda, K. Tokuda and H. Matsuda, J. Electroanal. Chem., 186 (1985) 79.
27. A. Szabo, D. K. Cope, D. E. Tallman, P. M. Kovach and R. M. Wightman, J. Electroanal. Chem., 217 (1987) 417.
28. A. Russell, K. Repka, T. Dibble, J. Ghoroghchian, J. Smith, M. Fleischmann and S. Pons, Anal. Chem., 58 (1986) 2961.
29. G. N. Watson, A Treatise on the Theory of Bessel Functions, 2nd Edition, Cambridge University Press, Cambridge (1948).
30. H. Gröber, Die Grundgesetzeder Wärmelectung und des Wärmeübergangls, Springer Verlag, Berlin (1921). See also the revised English edition: H. Gröber, S. Erk and V. Grigull, Fundamentals of Heat Transfer, McGraw-Hill, Inc., New York (1961).
31. H.S. Carslaw and J.C. Jaeger, Conduction of Heat in Solids, Clarendon Press, Oxford (1959).
32. Lord Kelvin, Reprints of Papers on Electrostatics and Magnetism, MacMillan, London (1872) p. 178.
33. J. C. Cooke, Q. J. Mech, Appl. Math., 16 (1963) 1.
34. A. Szabo, private communication and to be published.
35. W. R. Smythe, J. Appl. Phys., 22 (1951) 1499.
36. M. Fleischmann, F. Lasserre, J. Robinson and D. Swan, J. Electroanal. Chem., 177 (1984) 97.
37. M. Fleischmann, F. Lasserre and J. Robinson, J. Electroanal. Chem., 177 (1984) 115.
38. J.E.B. Randles, Disc. Faraday Soc., 1 (1947) 11.
39. M. Fleischmann and S. Pons, to be published.

3

Less is More:
Fabrication of Ultramicroelectrodes

Introduction

The size of a microelectrode has been redefined over time. In electrolytic and battery applications, 1cm^2 of electrode area once was considered a microelectrode (1). In electroanalytical applications (2), a microelectrode that has an area less than 0.1cm^2 is considered to be a standard, not small, size. The bulk of the literature citations on microelectrodes prior to 1984 reflects the usefulness of sensing or stimulating electrodes sized less than 10μm in electrophysiology and neuroscience. The recent emphasis among electrochemists interested in the electroanalytical, theoretical, and fundamental physical properties of electrodes with at least one dimension less than 1μm has given rise to the designation ultramicroelectrode. This survey of ultramicroelectrode fabrication methods is restricted to the construction of voltammetric electrodes with one dimension less than 1-10μm and, unfortunately, deliberately neglects the enormous body of work dealing with microelectrodes used *in vivo*. Some practical tips on electrode construction from this literature will be cited to acquaint both the new and practicing ultramicroelectrode fabricator with problems in common that have been observed and overcome.

The ultramicroelectrode jungle

Just as less of an electrode provides more electrochemical opportunities, fabricating less of an electrode is offset by greater difficulty in doing so. To date, ultramicroelectrodes appear in the following geometrical forms:

1. the microdisk, and its near relation, the microcylinder, followed by an akin concept;

2. the line electrode, which may stand alone or be assembled in an array;

3. the microring, which can be considered as a closed band electrode, or, for ease in mathematical description, the superposition of two unequal disks, one a sink and the other a source (see Theory chapter);

4. the microsphere and hemisphere;

5. irregular geometries — not necessarily fractal; and, of course,

6. other geometries, which can include colloids, conducting particle dispersions, and scanning tunnelling microscope tips.

Many of the ultramicroelectrodes are made by scaling down traditional electrode geometries, such as disks, wires, rings, and spheres, and rely on such traditional approaches as potting or sealing the conducting (or, less frequently, the semi-conducting) material in an insulating shroud. As the size of the wire or fiber drops below $10\mu m$ the sealing step becomes less straightforward and the quality of the electrode-to-insulator "bond" becomes crucial. Typically, a 100% success rate for ultramicroelectrode fabrication is neither achieved nor expected, whether using analogs of macrofabrication methods or using technologically more sophisticated approaches such as microlithographic patterning.

Table 1 lists the geometric forms of ultramicroelectrodes with general information on the material and means of

construction. Specific discussion on the construction of each
geometry will follow, but questions ultramicroelectrode users
should consider will be explored first. The intended use of the
ultramicroelectrode is the first filter on the choice of the
type of geometry; this will determine: (a) the actual size
tolerable in the application; (b) the most desirable electrode
material; (c) the compatibility of the insulating shroud or
support with respect to the electrode material and the stability
of the insulator in the electrolyte; (d) the necessity or
ability to reproducibly clean the electrode surface; (e) the
possibility of obtaining rigorous theoretical treatment of
current-potential curves for a given electrode geometry; (f) the
molecular turnover required; and (g) even the need or desire for
a technically simple method of construction.

The actual size of an ultramicroelectrode can be important
simply because not all ultramicroelectrodes as assembled are
small. A disk $1\mu m$ in diameter can be surrounded by 1-2mm of
glass or epoxy shroud. A band or line electrode may have a
submicrometer width, yet be centimeters long and surrounded by
several square centimeters of insulator. Specifying a
characteristic dimension of the ultramicroelectrode clarifies
this point. The characteristic dimension refers to the width of
exposed electrode material and typically is used in quantitative
or approximate treatments of the current response of the
ultramicroelectrode. For *in vivo* applications where the size of
the assembled electrode is critical, the dominance of microcylindrical and minimally shrouded microdisk geometries is
apparent (62).

The electrode material of choice for the electrochemical
application may not exist in a form that can be adapted to an
ultramicroelectrode shape, but with the variety of geometries
and multiplicity of construction approaches available this is

Table I: A Survey of

Ultramicroelectrode Geometry	Method of Construction	Electrode Material
I. Disk-Cylinder	1. Seal fine wires into an insulator	1a. Pt b. W c. Cu d. Ag e. Au f. Wollaston wire g. C fibers
	2. Electrodeposit Hg on a microcylindrical electrode	2. Hg on C
II. Disk-Cylinder Arrays	1. Grow a conducting polymer through a well defined porous insulator	1. Polypyrrole
	2. Anodize Al to form pores; metal coat the pore walls	2. Au
	3. Space fine wires 80-500 μm apart and set in epoxy	3a. C b. Pt
	4. Micromill holes through a resist coating a conductor or a conductor sandwich	4a. C b. Cr, Au
	5. Ag epoxy C fiber against a Cu rod; seal in a plastic tube; pot with epoxy	5. C, Hg on C
	6. Bundle C fibers and epoxy into the tip of a glass capillary	6. C, Hg on C
III. Spheres and Hemispheres	1. Dropping Hg electrode	1. Pure Hg
	2. Electrodeposit M on a microdisk	2. Hg on Pt
	3. All microdisks approximate a microhemisphere	
IV. Ring	1. Pyrolyze CH_4 inside of a drawn capillary	1. Pyrolytic C
	2. Focus a light beam on a rotating semiconductor disk	2. InP, GaAs
	3. Paint commercial metallo-organic ink on the exterior of a drawn glass fiber; fire to the conducting M; seal in glass	3. Au
	4. Sputter M onto a drawn capillary; seal in glass	4a. Pt b. Au
V. Tubular	1. Seal M foil between two insulating planes; drill through the sandwich	1a. Pt b. Pyrolytic C

Fabricated Ultramicroelectrodes

Insulating Material	Characteristic Dimension[1]/μm	Reference
glass, epoxy	0.1-0.3, more typically 1-5	1a. (3-9) b. (10-11) c,d. (11) e. (3,12);(39) f. (5-8) g. (13-21) 2. (22);(33)
1. Nucleopore	0.005-6	1. (23)
2. Al$_2$O$_3$		2. (24)
3. epoxy	3a. 5 b. 12.5	3a. (25) b. (26)
4.a. photoresist b. polymethylmethacrylate resist	4a. 10 b. 0.375	4a. (27) b. (28)
5. polyethylene, epoxy core	5. 4.6	5. (29)
6. epoxy	6. 4-5	6. (30,31)
1. none 2. glass	1. 0.05-10 2a. r_{Pt} = 15 b. r_{Pt} = 0.3; r_{Hg} = 2.3	1. (79) 2a. (32) b. (5)
1. glass, epoxy core	1. 0.012-0.12	1. (34)
2. unspecified	2. 100	2. (35)
3. glass	3. 0.09-0.9	3. (36,37)
4. glass	4a. 0.03-0.05 b. 0.4	4a. (38);(49) b. (49)
1. a. glass b. heat-bondable plastic to stainless steel	1a. 18 b. 3-6	1. (4)

1. Characteristic dimension is the radius of a cylinder, disk or sphere or the width of a line or ring electrode.

Table I: A Survey of

Ultramicroelectrode Geometry	Method of Construction	Electrode Material
VI. Band or Line	1. Sputter conducting film on insulator; epoxy to a glass slide; M exposed by grinding edge face 2. Seal M foil between glass slides; M exposed by grinding edge face 3. Silk screen metallo-organic ink on glass; fire to M; epoxy to a glass slide; edge face exposed	1a. Pt, Au b. Au on Mylar c. C on glass d. Pt on mica 2a. Pt b. Pyrolytic C 3. Au, Pt
VII. Line Arrays	1. Sputter M or MOx on insulator; pattern photolithographically 2. Electrodeposit Pt on Au line array 3. Vapor deposit M on glass; photolithographically pattern; epoxy on a cover glass; edge face ground flat 4. Silk screen metallo-organic ink on glass; fire to M; epoxy to a glass slide; edge face exposed	1a. Au b. Pt c. SnO_2, In_2O_3/Sn 2. Pt on Au 3. Au 4. Au, Pt
VIII. Irregular Geometries usually arrays	1. Impregnate porous conductor with epoxy 2. Form composites of conducting powders and insulator	1. reticulated vitreous C 2. graphite
IX. Spherical Dispersions	1. Disperse M particle in solvent, access between parallel feeder electrodes 2. Form M, MOx, or semiconducting colloids dispersed in solvent; access photolytically	1. Pt, Au, C 2. Pt, Au, TiOx RuOx, CdS,..
X. Other	1. M or MOx supported on alumina or zeolite; access as in IX#1 2. Tip of a scanning tunnelling microscope	1. Pt 2. Pt-Ir

1. Characteristic dimension is the radius of a cylinder, disk or sphere or the width of a line or ring electrode.

Fabricated Ultramicroelectrodes (Cont.)

Insulating Material	Characteristic Dimension[1]/μm	Reference
1. glass, epoxy	1a. 0.03-2.3 b. 0.005 c. 0.03 d. 0.01-6	1a-c. (38) a,b. (39) d. (61)
2a. glass b. heat-bondable plastic to stainless steel	2a. 18 b. 3	2. (4)
3. glass, epoxy	3. 0.05-0.2	3. (39)
1a. Si/SiO$_2$; quartz b. glass; polyimide c. glass	1a. 3/50-60 b. 3.5 c. 60	1a. (40-43) b. (44,45) c. (42)
2. Si/SiO$_2$	2. 1.5-4.1	2. (46,47)
3. glass, epoxy	3. 0.1	3. (48)
4. glass, epoxy	4. 0.05-0.2	4. (48)
1. epoxy	1. ~ 30	1. (50,51)
2a. Kel-F 2b. epoxy	2a. ~ 25	2a. (52) b. (53)
1. Not applicable	1. 0.1-1; 5-22	1. (54);(55)
2. Not applicable	2. 0.003-0.1	2. (56-58) and references therein
1. Al$_2$O$_3$; zeolite	1. 0.001-0.01	1. (59)
2. glass		2. (60,85,86)

usually a surmountable problem. Although carbon, gold and platinum are the preeminent ultramicroelectrode materials, any material which eventually will conduct and can be made in micrometer thick wires or foils, melted, vapor deposited, sputtered, electrodeposited, dispersed in a conducting paint, pyrolyzed, formed colloidally, sampled photoelectrically with micrometer light beams, or supported as metal clusters stands a chance of being transformed into an ultramicroelectrode.

Micro to macro transitions

After choosing the geometry and electrode material, the conductor must be transformed into a working representation of the geometry. To be useful, the very small must be connected to the macroscopic world for ease in handling and to make electrical contact. So, almost all ultramicroelectrodes are supported on or shrouded within an insulating material. Interfacial problems often arise between the conductor and the insulator (notated as M-I); these must be minimized for successful electrode performance. While this is true for shrouded or supported electrodes of any size, M-I compatibility on an ultramicrolevel is critical.

The conductor-glass interface

Conducting wires or fibers heat sealed in glass are sensitive to the match between the coefficients of thermal expansion for the particular M-I (63). Most metals and glasses are poorly matched in thermal expansion; severe mismatch leads to cracks at the M-I junction. Microcracks are a source of nonlinear diffusion (or edge effects) and leakage at macroscopic wires heat sealed in glass. An ultramicroelectrode is an electrode that is all edge (i.e., the perimeter to area ratio is large). Microcracks at the M-I junction of an ultramicroelectrode provide an apparent electrode area which dwarfs the electrode's geometric area.

Pt and soft glass have essentially equivalent linear coefficients of expansion; Au and soft glass do not (63). By measuring residual currents and calculating the double layer capacitance as a function of scan rate, Wehmeyer and Wightman (12) have tracked the quality of Au microdisk electrodes prepared by flame sealing 10μm Au wires into soft glass. The capacitance per unit area values were larger at the ultramicroelectrodes and only dropped to those of conventionally sized Au disk electrodes (sheathed in Teflon) at high scan rates. The greater apparent capacity (and area) at the ultramicroelectrodes was ascribed to defects in the Au-glass seal. Pt microdisk electrodes had comparable normalized capacitances to those of large Pt disk electrodes at all scan rates, as expected for the higher quality of a Pt-soft glass seal. Silanizing Au ultramicroelectrodes after construction reduced residual currents, presumably by preventing seepage of water into the silanized, and thus, hydrophobic cracks (12). This study shows that unavoidable M-I defects may be diminished chemically via silanization or instrumentally, by using the electrode at high scan rates. When feasible for the application, some workers have also electrodeposited Hg to fill microcracks with success (5,29).

Several other strategies exist to minimize the formation of microcracks when sealing glass around conductors. The first practical step is to draw a glass capillary down before inserting the wire or fiber or to purchase microcapillaries with inner diameters less than 2μm (64). This diminishes the gap that must be collapsed around the wire. A second step is to lower the heat applied directly to the wire and the glass capillary. Heat sealing under vacuum has been used, (6,7), as has resistive heating of a wire coiled around the glass capillary (9,65), which presumably allows for more localized heating of the glass (65).

Direct flame sealing in glass, especially of carbon fibers, can affect the electrochemical characteristics of the formed ultramicroelectrode. Golas and Osteryoung (22) found that $8\mu m$ carbon fibers that were flame sealed in glass while taking care to keep the fiber cool, nonetheless, were more fragile and required more electrochemical preconditioning than like fibers sealed to glass with heat-shrinkable plastic. Preconditioning insured that the flame-sealed microcylindrical electrodes gave comparable voltammetric responses, though, presumably they were still fragile.

The conductor-epoxy interface

To avoid the interfacial problems arising from heat sealing metal in glass, many users of ultramicroelectrodes simply cast the metal in an epoxy resin or use epoxy to bond the electrode material to a shrouding or support material. This can be far less technically demanding during preparation, but the use of an epoxy as a sealant or as a sheath raises other concerns. The epoxy must adhere well to the electrode material or the seal will leak and render the ultramicroelectrode useless. Adequate adhesion to the primary support or shrouding material is required also, but is not quite so critical. Based on most of the literature to date, distress of the metal-epoxy junction and its effect on an ultramicroelectrode's voltammetric behavior has not been perceived as a severe problem. However, some concern with the completeness of the carbon-epoxy resin seal has been raised (22,3).

The stability of the epoxy in the electrolyte needs to be assessed, especially with regard to switching between aqueous and nonaqueous media. While the metal-epoxy seal may not leak, species leached from the epoxy may degrade the quality of the voltammetric response and lead to an aged epoxy with microscopic defects. Thormann, van den Bosch, and Bond reported that as the amount of epoxy resin used to cement a cover plate over a linear array of Au microbands increased, so did the residual currents they measured in acetonitrile (48).

Now that nontraditional, nonpolar solvents can be explored using ultramicroelectrodes, the durability of an epoxy resin can be stressed even further. Table II lists some epoxy resins that have been used either to shroud or bond ultramicroelectrode materials and includes the electrolyte systems to which they have been exposed. Other reports of epoxy resins used to fabricate ultramicroelectrodes exist, but do not specify the epoxy used.

Adhesion is also an important factor when the electrode material is painted, sputtered, evaporated, or melted onto an insulating, frequently glass, support. Noble metals especially will not adhere well to glass due to a low value of the work of adhesion. Undercoatings are used to prevent the metal from floating off the support during experiments. Cr or Ti underlayers routinely are applied to glass before thin Au films are sputtered or evaporated on. The undercoat, however, may not act as an inert adhesive. Interdiffusion of Ti into the Au layer occurs with time; there is some concern that this phenomenon leads to contact resistance problems and eventual failure of the electrode (69). Interdiffusion of Cr to the surface of the Au adds Cr electrochemistry (27) to the residual response of the ultramicroelectrode.

For many voltammetric applications, Pt is the preferred electrode material (63,70). Vapor depositing durable films of Pt on glass is sufficiently troublesome that some groups rely on

Table II: Epoxy Resins Used in the

Epoxy Resin	Source	UME Material	Geometry
Sody 33 (polyester resin)	Escil	C	cylinder
Epon 828 + 14% m-phenylenediamine[2]	Miller-Stephenson (Chicago)	C	disk[3]
		C	disk
		Au	band
		C	cylinder
		C	disk
Epoxide resin + hardener	Buehler	RVC	irregular array
Epo-Tek 320	Epoxy Technology (Watertown, MA)	RVC, Pt	irregular array, disk array
5-min epoxy glue	Devcon	C	cylinder
Acculite cyanacryl glue[4]	Cotronics Corp.	C	disk, array, cylinder
Araldite[5]	Ciba-Geigy	Au, Pt	band array
353ND or 2106T[6]	Epoxy Technology, Inc. Tra-Con Corp.	C	cylinder
Epoxi-Patch	Dexter Corp., Hysol Division	Pt	disk RDE
		SnO_2	band RDE
		Au	band array
Apiezon vacuum wax	Apiezon	Pt	IDA
Torr-Seal vacuum epoxy	Varian	C	disk

1. Characteristic dimension is defined as the radius of the wire or the width of the band.
2. Wightman's group prefers Epon 828 because it is a widely used epoxy in carbon-fiber composite materials and because of its bonding strength and inertness to chemical attack (16).

Construction of Ultramicroelectrodes

char. dim.[1] μm	Electrolyte	Redox Species	Reference
4	phosphate buffer	catechols	(13)
2.5-5	1$\underline{M}$KCl	$Fe(CN)_6^{3-}$	(15)
≈ 7	LiClO$_4$/AN	ferrocene	(3)
	phosphate buffer KCl	$Ru(NH_3)_6^{3+}$ $Fe(CN)_6^{3-}$	(38)
5	phosphate buffer	$M(NH_3)_6^{n+}$, $Fe(CN)_6^{3-}$, catechols	(16)
0.012-0.12	1$\underline{M}$KCl	$Fe(CN)_6^{3-}$	(34)
	0.1$\underline{M}$KCl	$Fe(CN)_6^{4-}$	(50)
	0.1$\underline{M}$KNO$_3$	$Fe(CN)_6^{4-}$	(51)
<7	acetate buffer + EDTA	catechols, ascorbic acid	(18)
4-5	0.1$\underline{M}$HCl	M^{2+} ions	(30)
0.05-0.2	Et$_4$NClO$_4$/AN AN	ferrocene ferrocene	(48)
5.5	acetate buffer	$\mu\underline{M}$ $Ru(bpy)_3^{2+}$	(66)
12.5 0.25	0.4$\underline{M}$KCl, 4:1 H$_2$O:AN 1$\underline{M}$KCl	$Fe(CN)_6^{3-}$ $Ru(NH_3)_6^{3+}$	(67)
3	0.1$\underline{M}$LiClO$_4$ 0.1$\underline{M}$Bu$_4$NClO$_4$/AN	$Fe(CN)_6^{4-/3-}$ pyrrole	(40)
3.5	1$\underline{M}$H$_2$SO$_4$ 0.1$\underline{M}$Et$_4$NClO$_4$/AN 0.01$\underline{M}$KCl+0.01$\underline{M}$HCl	os(bpy)$_2$(vpy)$_2$Cl$_2$, Ru(bpy)$_3^{2+}$ $Fe(CN)_6^{3-}$+FeCl$_3$	(44)
2.5			(68)

3. Fiber + glass tip were dipped into the expoy and cut flush to the glass.
4. The electrodes were found to be leakage free.
5. The magnitude of the residual current increases as the amount of epoxy exposed to the solution increases.
6. No electrochemical pretreatment was reported.

electrodeposition of Pt on micropatterned Au arrays (46,47). Ti has been used by Chidsey and Murray as an adhesive underlayer for vapor-deposited Pt prior to micropatterning interdigitated arrays (44), while White, et al., have used a Ta coating between a biocompatible polymer and a thin film of Pt to fabricate a flexible thin-film microelectrode line array for use in prosthesis (45). Metallo-organic conducting inks adhere when fired because base metals present in the ink form adherent metal oxides which act as the noble metal-glass glue. Au and Pt band, microring, and linear array ultramicroelectrodes have been formed with good adherence on glass using the conducting inks (see Table I).

The other link to the macroscopic world is the ohmic contact to the ultramicroconductor. Often the route is covered in stages to avoid mechanical wear on the direct contact to the ultramicroconductor. Workers with carbon fibers often use Hg to contact the ensheathed fiber and a copper or nichrome wire for external connection. With metal wires, connection strategies include spot-welding or using conductive epoxies or silver solder. An epoxy plug has been used frequently to stabilize the contact point and the lead wire from vibration and from the wear and tear of mechanical attachment. To minimize vibrational noise and any possible triboelectric response (36), Fleischmann and co-workers silver soldered Pt wire ($r<2.5\mu$m) to a fine helical spring which in turn was soldered to a coaxial cable. The cable was epoxied into the top of the glass capillary to yield a rugged electrical contact (6,7).

Ultramicroelectrodes prepared with conducting paints and/or using microlithographic patterning are designed with a larger area of metal as a contact pad. This makes it relatively easy to form durable electrical attachments, although the contact area must then be masked to maintain a well defined working electrode area. Some use has been made of the wire-bonding and encapsulation technology routinely employed by the microelectronics industry (40), but this approach requires special equipment.

Cleaning/pretreating ultramicrosurfaces

By this point, the ultramicroelectrode's geometry and electrode material have been selected and a working example has been fabricated after consideration of the compatibility of the insulating shroud or support with everything it may contact and after a hardy electrical connection has been made. Now, this work of science, or technology or more likely art must be used electrochemically, so the usual concerns about the state of any electrode's surface must be considered. As is customary at conventional electrodes, polishing and chemical or electrochemical pretreatments perform most of the cleaning chores at ultramicroelectrode surfaces. An ultramicroelectrode surface capable of direct polishing is typically polished with successively finer grit sizes and finished with $0.05\mu m$ alumina. Recommendations for complete removal of alumina polishing compound include a final polish on a water-dampened polishing cloth (6); a wash with dilute nitric acid (71); a brief rinse with 95% EtOH, followed by ultrasonication in water (64); or a final polish on balsa wood, which is soft enough to preferentially hold the alumina particles (72). Wightman recommends that a final polish with diamond paste be avoided; irreproducible limiting currents were obtained on Au, Pt, and C microband electrodes, presumably due to the difficulty of complete removal of the paste (38).

Polishable ultramicroelectrode surfaces are commonly repolished with $0.05\mu m$ alumina prior to each day's use or as required. Bond and Fleischmann (6,7) report extended lifetimes for their Pt microdisk electrodes (in excess of a year or for hundreds of experiments) using this method. Ensheathed ultramicroelectrodes run the risk of becoming recessed into the insulator with excessive polishing. This allows a quiescent solution layer to form over the electrode surface reimposing linear diffusion to the surface. Wightman and co-workers observed that the limiting currents at Au, Pt, and C line ultramicroelectrodes decreased after extensive polishing and eventually assumed a peak shaped, diffusion-controlled response (38). Nonetheless, Wightman's group still repolishes their microdisk and microband electrodes with $0.05\mu m$ alumina before each use. Certainly, ultramicroelectrode surfaces benefit from a light touch in the polishing step.

Not all ultramicroelectrode geometries tolerate mechanical polishing; most prominent of these are the microcylindrical and microlithographic patterned examples. Some reports with these geometries include no information on surface pretreatment. A variety of chemical and electrochemical pretreatment procedures do exist for carbon microcylindrical electrodes and these are summarized in Table III. Pretreatment procedures for carbon fiber ultramicroelectrodes have evolved to the use of a combined AC signal-potentiostatic conditioning first proposed by Pujol in 1981 (73). Kovach, Deakin, and Wightman adapted Pujol's procedure and have recently evaluated the voltammetric characteristics of varied redox species at carbon fiber ultramicroelectrodes after electrochemical pretreatment (16). The treated fibers were microscopically roughened and cracked, in line with an observed fivefold increase in capacitance. The surface oxidation resulting from pretreatment creates cation-exchange sites leading to strong adsorption of dopamine and electroactive cations but, also, blocks the area available for electron transfer for nonadsorbed anionic redox species. This extreme transformation of carbon-fiber surfaces with oxidation may not be desirable for all applications, but it provides long term stability to carbon-fiber ultramicroelectrodes used for *in vivo* measurements (73,16).

Microelectrode arrays patterned by microlithographic techniques also must be cleaned via nonmechanical methods. In the early work with Au linear arrays, Wrighton's group relied on H_2 evolution in aqueous electrolyte to clean the Au surface (40,41). Due to the difficulty in obtaining reproducible electrochemical characteristics at the Au arrays, the arrays were converted to Pt by electrodeposition of Pt from aqueous 2m$\underline{M}$ K_2PtCl_6/0.1$\underline{M}$ K_2HPO_4. The electrodeposited Pt on Au arrays did yield reproducible responses (45,46). Before electrodeposition, the Au electrodes were cleaned; one of the latest variants (46) involved 5 cycles between -1.6 and -2.0V(SCE) in 0.1$\underline{M}$ K_2HPO_4 to evolve H_2. After electrodeposition, the Pt arrays were coated with electroactive polymers, which obviates the need to recondition the surface of the Pt electrode elements with use.

What you see . . .

Finally, after fabrication and surfacing, but before electrochemical use, the agreement between the ultramicroelectrode as envisioned and the ultramicroelectrode as prepared should be appraised. Osteryoung has emphasized the importance of optical microscopy as a routine check of the quality of the fabricated ultramicroelectrode (75). This is a convenient method to assess the regularity of the geometry (how perfectly circular is that 0.6μm disk?) and to survey any imperfections that may have arisen at the conductor-insulator junction during either fabrication or polishing. A true geometry is critical if the electrochemical response of the ultramicroelectrode will be compared to theoretical predictions (64).

Table III: Chemical and Electrochemical Pretreatment

Electrode Material	Radius/μm	Pretreatment Procedure
C	4	Normal pulse polarographic cycle from -0.8 to +1.2V(SCE) in phosphate-buffered saline.
C	4	Applied a 70 Hz alternating potential triangular wave from 0 to +3V (Ag/AgCl) for 20s followed by potentiostatting at +1.5V for 20s in a phosphate-buffered saline.
C[1]	4-6	Applied a variable amplitude sine wave (70Hz) from an initial potential of -0.3 to -1.1V (Ag/AgCl) in 0.1MKCl for < 3.5min; most fibers were treated for 0.5min.
C/Hg[2]	4-5	Used I_2/I^- to dissolve unwanted Hg.
C	<7	Potentiostatted for 1min at +1.3V(Ag/AgCl).
C	4	Potentiostatted for 10min at +0.7V(SCE) in 0.1MKSCN until current dropped below 0.1μm.
C	4	Potential swept from -0.2 to 2.0V (SCE) at 0.2V/s for 2 min in phosphate-buffered saline; potential swept from -0.2 to +0.8V at 5mV/s until a stable background current is obtained.
C	5	Applied a triangular wave (70Hz) from 0.0 to 3.0V (SSCE) for 20s; held at +1.5V for 20s; held at 0.0V for 30-60min -- in phosphate buffer (pH 7.4).
Pt	5	Treated with hot 1:1 HNO_3/H_2SO_4 and rinsed with water; electrochemically cycled between -0.1 and +1.0V (SCE) at 20V/s for ~ 6000 cycles; after transfer to the sample cell, the electrode is held at +0.2V for several seconds [74].

1. Single and bundles of C fibers were fabricated as microcylindrical electrodes.

Procedures for Cylindrical Ultramicroelectrodes

Observations	References
Lower residual currents were obtained for the second and subsequent sweeps.	(13)
Electrodes break at an applied upper E>3.6V, but the oxidation of ascorbate is not separated from that of catechols at an applied upper E<2.7V.	(73)
>3.5min of pretreatment caused many fibers to break or produced erratic electrochemistry; Avoid potentiostatting--less damage with AC signal; greatest improvement in sensitivity occurred by cycling into the oxygen wave and avoiding the hydrogen wave; a concomitant increase in ionizable surface groups also results.	(30)
Strong oxidants, e.g., HNO_3, yield irreproducible electrode behavior; Potentiostatting at positive potentials destroyed some C fibers.	(29)
	(18)
	(22)
Well defined, reproducible voltammetry for dopamine is observed if pretreatment scan is extended to 2.0V rather than 1.5V	(78)
The final potentiostatting step at 0.0V is required to obtain a more stable electrochemical response.	(16)
Reproducible electrochemical responses resulted; no surface waves were observed in the potential window used for voltammetry.	(4)

2. Sharper and more symmetrical peaks were obtained for the C-fiber microcylindrical arrays after the pretreatment; this was attributed to an increase in surface area and to an increase in the number of quinoidal groups at the C surface.

Thormann and Bond have recently explored transient voltammetric techniques as a diagnostic of the fabrication quality of ultramicroelectrodes (76). They note, as have other researchers, that at short times nonideal influences on the voltammetric signal are magnified and this is when imperfect construction of the ultramicroelectrode will be most telling. Square wave, differential pulse, and AC voltammetry and short-time chronocoulometry were applied to Au and Pt line and ring ultramicroelectrodes in 1m$\underline{M}$ ferrocene/CH_3CN.

Imperfections in the metal to insulator junction were apparent using any of the techniques at short times, and were observed most readily at epoxy-sealed electrodes where the thickness of the epoxy layer around the conductor was larger than the characteristic dimension of the conductor. Glass or epoxy-sealed ultramicroelectrodes without cracks gave results consistent with theoretical predictions for transient voltammetric techniques applied to ultramicroelectrodes.

The shape of the chronocoulogram gives a rapid, qualitative evaluation of the diffusional conditions at the ultramicroelectrode. Ultramicroelectrodes without imperfections exhibit a chronocoulogram that is linear during the pulse followed by a time-independent charge; this is the expected result for an electrode under steady state or quasi steady state mass transfer conditions, as an ultramicroelectrode should be. The goodness of the fit of the metal to the insulator can be quickly determined with short-time chronocoulometry, while the transient voltammetric techniques can be used to rigorously assess theoretical concurrence.

How much less???

How much smaller can ultramicroelectrodes be made? Is the ultimate voltaic microelectrode (one approaching atomic scales) feasible? In several senses, the answer is yes, and the future is here.

"Wireless" electrochemistry

Dispersions of colloidal metals and semiconductors behave as microelectrodes, as pointed out by Henglein in 1980 (82), but electrochemistry at these colloids is traditionally driven with light (57,58). A spherical 2-nm wide Pt particle contains 277 Pt atoms; a spherical 1-nm wide Pt particle contains 35 Pt atoms. Colloidal particles sized at 2nm are not uncommon and zeolites can be used to synthetically template 1nm or smaller metal or metal oxide particles (83).

How can such atomic electrodes be harnessed without relying solely on photoelectrochemistry? Work by Pons, Fleischmann, et al., with $0.1\mu m$ Pt and Au microspheres has shown that bipolar electrolysis occurs when these small spheres are dispersed between feeder electrodes (54); this technique permits "wireless" electrical access to the microspheres. During dispersion electrolysis, the microspheres retain advantages associated with ultramicroelectrodes, e.g., the ability to be used in solvents without deliberately added electrolyte salt, yet provide multiple microelectrode sites for practical electrolytic results (84,54).

Dispersion electrolysis is a valid approach as well for smaller Pt particles supported in/on zeolite Type Y (59). In addition to the template nature of the zeolite, the catalytic and molecular sieving character of this support is expected to influence susceptible electrode processes by affecting the chemical step in CE or EC reactions. Dispersion electrolysis offers the means to explore some of the smallest of electrodes.

Electrochemical microscopy

Recent work with scanning tunneling microscopy (STM) has discussed and demonstrated (60,85,86) the promise of this technique in liquids suitable for electrochemical studies. Features on the surface of graphite (60) and an integrated circuit (85) were tracked in pure water using STM.

Rather than rely on the tunneling current flowing between the STM tip and the surface being scanned, the tip can be backed off slightly from the 1nm distance used for tunneling measurements and faradaic current can be imposed. Bard and co-workers have styled this as scanning electrochemical and tunneling microscopy (SETM) (85,86).

The STM or SETM tip behaves as an ultramicroelectrode and effective mass transport to the tip results. Bard et al., combined these features to electrochemically etch submicrometer patterns on illuminated GaAs (86). The combination of STM and SETM enables reading of the (macro)electrode's surface features via STM and writing on the (macro)electrode's surface with the electrochemical etching, deposition, or modification reactions possible via SETM.

Indeed, less electrode offers more electrochemical opportunities. The indications are that even less will eventually be even more.

References

1. apocryphal
2. A.J. Bard and L.R. Faulkner, "Electrochemical Methods", Wiley and Sons, New York, 1980, Ch. 5-7.
3. J.O. Howell and R.M. Wightman, Anal. Chem., 56 (1984) 524-529.
4. P.M. Kovach, W.L. Caudill, D.G. Peters and R.M. Wightman, J. Electroanal. Chem., 185 (1985) 285-295.
5. K.R. Wehmeyer and R.M. Wightman, Anal. Chem., 57 (1985) 1989-1993.
6. A.M. Bond, M. Fleischmann and J. Robinson, J. Electroanal. Chem., 168 (1984) 299-312.
7. M. Fleischmann, F. Lasserre, J. Robinson and D. Swan, J. Electroanal. Chem., 177 (1984) 97-114.
8. B.J. Feldman, A.G. Ewing and R.W. Murray, J. Electroanal. Chem., 194 (1985) 63-81.
9. L.O. Whiteley and C.R. Martin, J. Electroanal. Chem., in the press.
10. J.C Nussbaumer, J. Neurosci. Methods 3 (1981) 247-250.
11. D. Swan, Ph.D. Thesis, The University, Southampton (1981).
12. K.R. Wehmeyer and R.M. Wightman, J. Electroanal. Chem., 196 (1985) 417-421.
13. J.L. Ponchon, R. Cespuglio, F. Gonon, M. Jouvet and J.F. Pujol, Anal. Chem., 51 (1979) 1483-1486.
14. N. Armstrong-James and J. Miller, J. Neurosci. Methods, 1 (1979) 279-287.
15. M.A. Dayton, J.C. Brown, K.J. Stutts and R.M. Wightman, Anal. Chem., 52 (1980) 946-950.
16. P.M. Kovach, M.R. Deakin and R.M. Wightman, J. Phys. Chem., 90 (1986) 4612-4617.
17. C.W. Anderson and M.R. Cushman, J. Neurosci. Methods, 4 (1981) 435-436.
18. L.A. Knecht, E.J. Guthrie and J.W. Jorgenson, Anal. Chem., 56 (1984) 479-482.
19. K. Aoki, K. Honda, K. Tokuda and H. Matsuda, J. Electroanal. Chem., 186 (1985) 79-86.
20. A.S. Baranski, J. Electrochem. Soc., 133 (1986) 93-97.
21. J.W. Bixler and A.M. Bond, manuscript.
22. J. Golas and J. Osteryoung, Anal. Chim. Acta, 181 (1986) 211-218.
23. R.M. Penner and C.R. Martin, J. Electrochem. Soc., 133 (1986) 2206-2207.
24. C.J. Miller and M. Majda, J. Electroanal. Chem., 207 (1986) 49-72.
25. W.L. Caudill, J.O. Howell and R.M. Wightman, Anal. Chem., 54 (1982) 2532-2535.
26. R.C. Engstrom, Anal. Chem., 56 (1984) 890-894.

27. K. Aoki and J. Osteryoung, J. Electroanal. Chem., 125 (1981) 315-320.
28. T. Hepel and J. Osteryoung, J. Electrochim. Soc., 133 (1986) 752-757.
29. M. Ciszkowska and Z. Stojek, J. Electroanal. Chem., 191 (1985) 101-110.
30. G. Schulze and W. Frenzel, Anal. Chim. Acta, 159 (1984) 95-103.
31. T.E. Edmonds and J. Guoliang, Anal. Chim. Acta, 151 (1983) 99-108.
32. R. Lines and V.D. Parker, Acta Chemica Scand., B31 (1977) 369-374.
33. M.R. Cushman, B.G. Bennett, and C.W. Anderson, Anal. Chim. Acta, 130 (1981) 323-327.
34. Y.T. Kim, D.M. Scarnulio, and A.G. Ewing, Anal. Chem., 58 (1986) 1782-1786.
35. J.M. Rosamilia and B. Miller, J. Electrochem. Soc., 132 (1985) 2621-2626.
36. A. Russell, K. Repka, T. Dibble, J. Ghoroghchian, J.J. Smith, M. Fleischmann and S. Pons, Anal. Chem., 58 (1986) 2961-2964.
37. T. Dibble, S. Bandyopadhyay, J. Ghoroghchian, J.J. Smith, F. Sarfarazi, M. Fleischmann, and S. Pons, J. Phys. Chem., 90 (1986) 5275-5277.
38. K.R. Wehmeyer, M.R. Deakin, and R.M. Wightman, Anal. Chem., 57 (1985) 1913-1916.
39. A.M. Bond, T.L.E. Henderson, and W. Thormann, J. Phys. Chem., 90 (1986) 2911-2917.
40. G.P. Kittlesen, H.S. White, and M.S. Wrighton, J. Am. Chem. Soc., 106 (1984) 7389-7396.
41. J.W. Thackeray, H.S. White, and M.S. Wrighton, J. Phys. Chem., 89 (1985) 5133-5140.
42. W. Thormann, D. Arn, and E. Schumacher, Separation Sci. and Tech., 19 (1984-1985) 995-1011.
43. D.G. Sanderson and L.B. Anderson, Anal. Chem., 57 (1985) 2388-2393.
44. C. Chidsey, B.J. Feldman, C. Lundgren, and R.W. Murray, Anal. Chem., 58 (1986) 601-607.
45. S.A. Shamma-Donoghue, G.A. May, N.E. Cotter, R.L. White, and F.B. Simmons, IEEE Trans. Electron Devices, 29 (1982) 136-144.
46. G.P. Kittlesen, H.S. White, and M.S. Wightman, J. Am. Chem. Soc., 107 (1985) 7373-7380.
47. A.J. Bard, J.A. Crayston, G.P. Kittlesen, T.V. Shea, and M.S. Wrighton, Anal. Chem., 58 (1986) 2321-2331.
48. W. Thormann, P. van den Bosch, and A.M. Bond, Anal. Chem., 57 (1985) 2764-2770.

49. D.R. MacFarlane and D.K.Y. Wong, J. Electroanal. Chem., **185** (1985) 197-202.
50. N. Sleszynski, J. Osteryoung, and M. Carter, Anal. Chem., **56** (1984) 130-135.
51. R.C. Engstrom, Anal. Chem., **56** (1984) 890-894.
52. D.E. Weisshaar and D.E. Tallman, Anal. Chem., **55** (1983) 1146-1151.
53. R.C. Engstrom, M. Weber, and J. Werth, Anal. Chem., **57** (1985) 933-936.
54. M. Fleischmann, J. Ghoroghchian, D. Rolison, and S. Pons, J. Phys. Chem., **90** (1986) 6392-6400.
55. B. Kastening and S. Spinzig, J. Electroanal. Chem., **214** (1986) 295-302.
56. A. Henglein, B. Lindig, and J. Westerhausen, Radiat. Phys. Chem., **23** (1984) 199-205.
57. A. Henglein, "Photochemical Conversion and Storage of Solar Energy, Part A", J. Rabani, ed., Weizmann Science Press, Jerusalem, 1982, 15-213.
58. M. Graetzel, Acc. Chem. Res., **14** (1981) 376-384.
59. D.R. Rolison, R.J. Nowak, J. Ghoroghchian, S. Pons, and M. Fleischmann, "Proceedings of The Third International Symposium on Molecular Electronic Devices", F. Carter and H. Wohltjen, eds., Elsevier, Amsterdam, 1987.
60. R. Sonnenfeld and P. Hansma, Science, **232** (1986) 211-213.
61. T.V. Shea and A.J. Bard, Anal. Chem., **59** (1987) 0000.
62. R.D. Purves, "Microelectrode Methods for Intracellular Recording", Academic Press, New York, 1981.
63. D.T. Sawyer and J.L. Roberts, Jr., "Experimental Electrochemistry for Chemists", Wiley, New York, 1974, 86-88.
64. D.P. Whelan, J.J. O'Dea, J. Osteryoung, K. Aoki, J. Electroanal. Chem., **202** (1986) 23-36.
65. G. Taylor and H.H.J. Girault, J. Electroanal. Chem., **208** (1986) 179-183.
66. W.F. Berry and S.G. Weber, J. Electroanal. Chem., **208** (1986) 77-84.
67. T.E. Mallouk, V. Cammarata, J.A. Crayston, and M.S. Wrighton, J. Phys. Chem., **90** (1986) 2150-2156.
68. J.W. Bixler and A.M. Bond, Anal. Chem., **59** (1987) 0000.
69. J. Janata, ONR-NSF Workshop on Ultramicroelectrodes, Homestead, UT, January 1986.
70. R.N. Adams, "Electrochemistry at Solid Electrodes", Marcel Dekker, New York, 1969.
71. B. Scharifker and G. Hills, J. Electroanal. Chem., **130** (1981) 81-97.
72. S. Pons, ONR-NSF Workshop on Ultramicroelectrodes, Homestead, UT, January 1986.

73. F. Gonon, C.M. Fombarlat, M.J. Buda, and J.F. Pujol, Anal. Chem., 53 (1981) 1386-1389.
74. S.W. Barr, K.L. Guyer, and M.J. Weaver, J. Electroanal. Chem., 111 (1980) 41-59.
75. J. Osteryoung, ONR-NSF Workshop on Ultramicroelectrodes, Homestead, UT, January 1986.
76. W. Thormann and A.M. Bond, J. Electroanal. Chem., 218 (1987) 187-196.
77. J. Pons and S. Pons, private communication.
78. S. Sujaritvanichpong, K. Aoki, K. Tokuda, and H. Matsuda, J. Electroanal. Chem., 198 (1986) 195-203.
79. S. Pons, M. Fleischmann, J. Pons, J. Daschbach, J. Electroanal. Chem., submitted.
80. D. Belanger and M.S. Wrighton, Anal. Chem., 59 (1987) 0000.
81. C.E.D. Chidsey and R.W. Murray, Science, 231 (1986) 25-31.
82. A. Henglein, J. Phys. Chem., 84 (1980) 3461-3467.
83. *Metal Microstructures in Zeolites*, P.A. Jacobs, N.I. Jaeger, P. Jiru, G. Schulz-Ekloff, eds., Studies in Surface Science and Catalysis, 12, Elsevier Scientific, Amsterdam, 1982.
84. M. Fleischmann, J. Ghoroghchian and S. Pons, J. Phys. Chem., 89 (1985) 5530-5536.
85. H-Y. Liu, F-R.R. Fan, C.W. Lin, and A.J. Bard, J. Am. Chem. Soc., 108 (1986) 3838-3839.
86. C.W. Lin, F-R.F. Fan, and A.J. Bard, J. Electrochem. Soc., 134 (1987) 1038-1039.

APPENDIX
DETAILS ON THE CONSTRUCTION OF SPECIFIC ULTRAMICROELECTRODE GEOMETRIES

This appendix will serve as a supplement to Table I; more detailed information on fabrication methods will be given for the geometries listed in Table I. Procedural variations will be noted if they differ in kind and not degree. Because the greatest construction difficulties arise for characteristic dimensions below 1μm, published methods will be emphasized that discuss successfully prepared submicrometer electrodes. New fabrication approaches that have not yet generated submicrometer electrodes, but may do so with refinement, will also be included.

Mercury microspheres

Fleischmann, Pons, et al. have developed a dropping mercury ultramicroelectrode (79). The device is constructed from a piece of thick walled capillary tubing containing a bulb for the mercury reservoir. A wire contact to the mercury is made at one end, while the other end is drawn down to an inside diameter < 1μm after the device is filled with mercury. Mercury droplets are driven from the device by a small controlled temperature oven.

Microcylinder - microdisk

Wires or fibers of r>5μm seem relatively straightforward to ensheathe after working with those of r<5μm. The procedures below are viable for fibers of r<5μm and can be used with fibers of r>5μm--the reverse could not necessarily be said.

Platinum

Sealing Wollaston wire in glass: Pt microdisks of r = 0.3 - 1µm (6,7)

1. A 1-cm length of Wollaston wire [Goodfellow Metals, Cambridge, England] is silver soldered to one end of a fine spring; the spring is soldered to the central connector of a low noise coaxial cable.

2. The wire/spring/cable is positioned inside of a pre-drawn Pyrex capillary, which contains a side arm, with the wire filling the capillary tip.

3. The coaxial cable is cemented with epoxy at the top of the capillary.

4. The silver cladding is dissolved by immersing the capillary tip in 50% HNO_3 for 4h, positioned so that only the lower half of the wire is exposed {a}.

5. The inside of the capillary tip is washed thoroughly with water added via the sidearm, rinsed with acetone, and left to dry.

6. The wire is sealed under vacuum (pulled through the sidearm) by heating gently - it is essential that the junction between the etched and unetched portions of the wire also be sealed or the wire will break upon polishing.

7. The cross-section of the wire is exposed by cutting the capillary tip and polishing with emery through successively finer grades of alumina.

8. The failure rate for 0.3µm disk electrodes is high {b}, but once made, an electrode can be used with repolishing for hundreds of experiments.

   {a} Wehmeyer and Wightman (5) have sealed Wollaston wires (r=0.3,1µm) in predrawn glass capillaries (i.d.≈1mm) after first removing the Ag coating.

   {b} Pons and Pons (77) have recently achieved r=0.05µm by etching the tip of a 25µm Pt wire in aqua regia; this approach has the great advantage of permitting one to work with a wire that remains thick at one end easing the handling and sealing tasks. A summary of the procedure follows.

1. 25μm Pt wire is sealed in soft glass with 7-8cm extending from the tip.

2. The lower 2.5cm of the wire is positioned in aqua regia at 100°C for 1h; the tip will be ≈100nm, while the body of the wire will be >15μm; rinse with water.

3. >3cm of the etched wire is clipped and the thick end inserted in a predrawn soft glass tube (o.d.≈6mm) with the fine end of the wire closely aligned to the tip of the capillary.

4. The seal is made with a microtorch which uses a 10 gauge hypodermic needle as the tip.

5. Ohmic contact is made by placing clean Cu wire in the tube without touching the fragile Pt wire; the wire is taped in place. A length of lead solder is placed next to the Cu and the lower end of the glass capillary (but not the tip) is heated for several seconds in a fine Bunsen flame to form a pool of solder encompassing the Pt and Cu wires.

6. The disk is exposed by wiping the tip with a soft cloth; major polishing of the tip has been found to be only occasionally necessary.

Carbon

Sealing fine C fibers into glass with and without epoxy seals:

C microcylinders of $r=5\mu$ (16)

1. A C fiber [Thornell VSB-32, Union Carbide] is aspirated into a glass capillary; the capillary is pulled to a fine taper with a pipet puller [Narishige Scientific Instrument Co., Tokyo, Japan].

2. The capillary is plugged by placing a second capillary filled with epoxy [Epon 828 + 14%m-phenylenediamine, Miller-Stephenson, Danbury, CT] at the fiber-glass boundary, so that the epoxy enters the electrode capillary by capillary action without coating the fiber; the epoxy is cured at 70°C.

3. The fiber is trimmed with a scalpel to a length of ≈500μm.

4. Electrical contact is made by filling the capillary with Hg and inserting a short wire.

C microcylinders of $r=4\mu m$ (22)

1. A 2.5-mm length of microbore Tygon tube (i.d.=0.4mm) is placed inside one end of a heat-shrinkable plastic tube (7-8mm long, i.d.=2mm). The other end of the heat-shrinkable tube is slipped onto the tip of a cut and partly sealed disposable glass pipet.

2. The C fiber [Aesar] is threaded into the capillary through the Tygon tube.

3. The Tygon tube is melted and the heat-shrinkable tube collapsed by gentle heating with a soldering iron; the tip should be observed with a magnifying glass to avoid overheating.

4. Ohmic contact is made with Pt or Ag wire to Hg.

C microdisks of $r=3.5\mu m$ (20)

1. 2-3 cm of the C fiber [Thornell 300, Union Carbide] is connected into a bored thin copper wire with silver epoxy and inserted into a polyethylene tube (i.d.=0.52mm).

2. The polyethylene tube is wrapped tightly with Teflon tape and Al foil and then heated to $\approx 200°C$ under vacuum. The polyethylene melts and completely seals the fiber in plastic.

3. Before use the polyethylene tube is cut cross-sectionally to expose a fresh surface of C; this step is repeated if the C surface is contaminated during use.

Microdisk - microcylinder arrays

Approaches to construct arrays of disk or cylinder ultramicroelectrodes vary from the usual case of sealing or potting a wire as occurs for single disk or cylinder microelectrodes. Patient multiple spacing and epoxying of single wires to form an array has been done, but several multiplex approaches will be highlighted here instead.

Polypyrrole

Electroforming a conducting polymer through a template: arrays of disk or cylindrical microelectrodes of r=5nm-6μm (23)

1. A Nucleopore membrane with a pore size of the desired diameter is overlaid on a planar Pt electrode.

2. Pyrrole is electropolymerized through the Nucleopore pores; the diameter of the electroformed polypyrrole fibrils is determined by the pore diameter of the Nucleopore membrane.

3. The Nucleopore membrane readily dissolves in CH_2Cl_2 leaving an array of conducting polymer fibers supported on Pt, i.e., an array of cylindrical ultramicroelectrodes. If this porous array is perfused with an insulating polymer or the Nucleopore membrane is retained, an array of ultramicro-electrode disks results.

Cr, Au

Milling submicrometer holes through a patterned resist coating a conductor or conductor sandwich: arrays of Cr or Au disk microelectrodes of r=375nm (28)

1. A thin layer of Cr and a 1-μm thick layer of Au followed by a thin overlayer of Cr are vapor deposited on a Si wafer. The under and overlayers of Cr act as an adhesive for the Au to the Si and for the photoresist, respectively. The Au layer acts as the current path so that all electrode elements are connected in parallel.

2. After coating with a 1-μm thick layer of polymethacrylate resist, the electrode pattern (a hexagonal configuration of disks of r=375nm spaced 10μm from a nearest neighbor) is imposed with a computer-controlled electron-beam apparatus. The initial electrode surface (Cr) is exposed after the beam etching; to expose a Au microelectrode array, the Cr is electrolytically dissolved.

3. The photoresist layer need not be removed; if not, upon use a 1-μm deep cylindrical well of solution covers each ultramicroelectrode reimposing a diffusional path from bulk solution to the ultramicroelectrode. This feature can be exploited to explore processes, such as corrosion, with and without diffusion control at the structured electrode array.

Microring

The ring geometry has been prepared routinely with some of the smallest characteristic dimensions to date. Many of the examples rely on coating the conductor on the walls of the insulating support, which often is a drawn glass capillary. Consequently, any eccentricities in the glass surface are reflected in the formed ring ultramicroelectrode. Exposing the ring through sanding or cutting can also introduce nonuniformities. In recent work (35), a light beam focussed on a rotating semiconductor disk electrode was used to generate a ring microelectrode with a large (100μm) characteristic dimension. Refinement of this approach offers ready access to semiconductors as microelectrodes.

Au
Sealing a thin layer of metal coated on a glass capillary in an insulator: Ring with a diameter of 10μm and a width of 90nm (36)

1. A 2mm quartz rod is flame drawn to make a fine fiber attached to the thicker rod.
2. The fiber/rod (a) is painted with a metallo-organic gold ink [Englehard] (b) with a thicker layer running up the rod; the painted rod is baked at 500°C for 15min to form an adherent Au(0) coating.
3. Electrical contact is made by soldering a wirewrap wire to the thick Au layer at the wider end of the rod.
4. The Au-coated rod can be sealed into a pipet or capillary tube with epoxy resin or by collapsing the glass around the rod.
5. The ring is exposed by cutting or sanding the end of the sealed assembly.
    (a) The converse of this procedure is to coat the inner walls of a fine capillary with the ink. The ring is formed after collapsing the fired capillary around a glass fiber or plugging the core with epoxy.

(b) The thin metal layer may be vapor deposited (38,49).

C
Pyrolyzing methane in a glass capillary to form pyrolytic C: ring with a width of 12-120nm (34)

1. A 1.3mm quartz capillary is pulled to a 1-4μm tip; the pulled capillary is placed in a 7mm quartz tube to prevent softening of the capillary upon heating.

2. CH_4 flowing through the capillary is pyrolyzed by applying the flame of a Bunsen burner to the exterior of the quartz tube.

3. The tip of the capillary is filled with epoxy [Epon 828 + 14% m-phenylenediamine].

4. The tip is cut with a scalpel; electrical contact is made via Hg and a nichrome wire.

Because of problems with ring uniformity, a more elaborate, i.e., a less routine construction approach has been devised which uses sputter coating to deposit the metal and a microtome to section the assembly to expose an undistorted ring. This approach also permits reproducible control of the ring dimensions (diameter and width). The procedure (79) is summarized below.

1. A blank is molded from a structurally strong, but electrically insulating material which can be easily cut using conventional electron microscopy microtomes. Suitable materials include fiberglass resins, Epon 828/m-phenylenediamine epoxy, and other electron microscopy epoxies. A cylinder with a nipple on the end is a successful shape for the blank.

2. A slit along the radius of the cylinder is made with a diamond saw through the length of the cylinder/nipple; a diamond saw is used to minimize the width of the slit.

3. A thin fiber with a diameter equal to the desired inside diameter of the microring is laid in the slit so that it is

centered through the cylinder/nipple and extends beyond the nipple; the fiber must also be of a material suitable for sectioning with a microtome. The slit is then filled with epoxy.

4. A groove is rasped into the outer wall of the cylinder with a triangular file. A 2.5cm piece of wire for electrical contact is laid in the groove and secured at the top and base of the cylinder with silver epoxy. The wire needs to be very clean; cleaning in acid is recommended.

5. The metal of choice for the ring is sputter coated on the modified blank. The depth of the deposit determines the width of the ring. Sputter coating deposits metal over the fiber, which will be the eventual electrode tip, the walls of the cylinder, and the lead wire; this ensures a continuous electrical connection from the tip of the electrode to the lead wire.

6. The blank is carefully inserted into a mold to avoid bending the fragile fiber and the mold is filled with insulating epoxy. After curing, the ensheathed blank is removed from the mold.

7. To expose the ring, the electrode is mounted into a microtome chuck and sectioned until the metal ring is exposed. Careful sectioning yields a smooth and uniform ring surface which needs no polishing before use. Sectioning with a glass knife also yields an undistorted ring. If the surface is fouled with use, a thin section can be made to expose a clean ring. This microelectrode form can be easily mounted for electron microscopy to obtain a precise determination of the dimensions of the ring.

Band or Line

Line electrodes offer a simple approach to electrodes with a very small width, so that ultramicroelectrochemical characteristics are retained, yet a large effective electrode area results due to the length of the band or line. This yields larger currents and assists the analytical usefulness of such ultramicroelectrodes.

The early line ultramicroelectrodes relied on sealing a thin sheet of the conductor (typically Pt, Au, or C) between two supports. The mechanical fragility of such thin foils limits the ability to prepare submicrometer electrodes; e.g., a thin sheet peeled from basal plane pyrolytic graphite eventually yielded a band electrode with a width ranging from 3-6μm (4), while even thicker Pt and Au foils were required. A more feasible approach for submicrometer line electrodes uses thin coatings of a conductor, as detailed below.

Microtubular electrodes have been prepared by sandwiching a metal or carbon layer between two insulating planes and then drilling through the sandwich to expose the electrode (4). Polishing of the surface is achieved by pneumatically forcing an alumina suspension through the hole.

Au, Pt
Sputtering thin metal films on glass: Au or Pt line ultramicroelectrodes of 30nm-2.3μm width (38)

1. A glass microscope slide is cleaned with detergent and water, oven dried, cleaned in 2-propanol and redried. The slide is masked with adhesive tape to expose a 0.6-1cm strip in the center.

2. Pt or Au is sputtered on the masked slide; after careful removal of the tape, the slides are heated at 200°C for ~2h.

3. A cover glass slip is epoxied [Epon 828 + 14% m-phenylenediamine] over part of the conductor (a). The glass slip is aligned so that one edge of the sandwich forms the band electrode; electrical contact is made at the other end by silver soldering a wire to the exposed metal film. The contact and the wire are masked with silicon rubber.

4. The band is exposed by grinding with successively finer grades of abrasive.

    (a) In addition to preparing sputtered films, Wightman's group also uses commercial sources of supported films. Au on Mylar [Goodfellow Metals, Cambridge, UK] is similarly sealed between two glass slides. Nominally 1-nm thick Au films on Mylar were found to be nonconductive; 5nm films are usable, as are 30nm carbon films on glass slides [Lebow, Goleta, CA]. Thormann (42,48), Bond (39), and co-workers form thin metal films using metallo-organic inks or vapor deposition of metals to prepare line and linear array ultramicroelectrodes; this work will be discussed in the section on line arrays.

Line Arrays

The use of photolithographic or silk-screening patterns with micrometer-sized lines has provided researchers with electrochemical devices which have immediate analytical and electronic applications.

Thormann has patterned arrays of vapor-deposited Au, SnO_2, or In_2O_3/SnO_2 with 256 60-μm wide lines spaced 340μm apart (42). These arrays are then used as isotachophoretic detectors. The contact pad area is designed so that each microelectrode is individually addressable. A complete multichannel scan can be made mechanically by tracking a graphite brush across the contact ends of the electrode elements or one microelectrode finger can be monitored as a function of time.

Wrighton, et. al., have explored arrays of individually addressable Au or Pt microelectrodes coated with conducting or electroactive polymers as electronic devices analogous to diodes and transistors (40,41,46, and references therein). Although such molecule-based electronic devices are slower than their solid state counterparts, the built-in signal amplification of these devices invites investigation of their capabilities as chemical sensors (80).

Murray (44,81) and Anderson (43) have demonstrated the analytical advantages of linear arrays that are interdigitated. In these arrays the pattern consists of one set of line ultramicroelectrodes connected in common to a contact pad alternately interlaced with an opposing set of line ultramicroelectrodes connected in common to a second contact pad. This design can be used as the microelectrode analog of twin-electrode analyses.

Anderson used his interdigitated electrode design in a thin-layer cell (43) and obtained large (due to the extended total electrode area) steady state (due to the microelectrode width) currents between the twin microelectrode elements of the array.

Murray and co-workers have exploited the ability to achieve steady state conditions at interdigitated arrays as a means to obtain a direct measure of the electron conduction in electroactive polymers formed over the array (44). Unlike other steady state approaches at conventionally sized modified electrodes, a simple expression was derived for D_e, the electron diffusion coefficient, which did not require knowledge of either the concentration of redox sites in the polymer film or the film thickness. Determinations or estimations of film swelling are, thus, unnecessary; this is a decided advantage of the interdigitated array over other steady state experiments. This expression for D_e, while dependent on the spacing between the twin electrode elements and the width of an electrode element, is independent of the length of the line ultramicroelectrode; thus, precise masking of the contact pad region is not required for significant measurements of D_e.

By insulating the face of a microelectrode array and exposing an edge face, Thormann, et. al., have easily prepared linear arrays of submicrometer electrodes (48). The characteristic dimension is now not the width of the line ($\gg 1 \mu m$), but the depth of the metal layer (typically on the order of $0.1 \mu m$). The fabrication method for this type of array and for the microlithographed examples will be detailed below.

Au, Pt

Sandwiching multi-element linear microelectrodes between insulators: linear ultramicroelectrode arrays with a characteristic dimension of 0.05-0.2µm (48)

1. A pattern of parallel lines is formed on thin Au or Pt layers by chemical etching (Au) or metallo-organic inks (Au or Pt). The specific procedures are outlined below.

    A. Photolithographic fabrication

    i. A 0.1-µm thick layer of Au is vapor deposited on 1-2mm thick glass plate. Some, but not all, samples were prior coated with a thin layer of Cr (a).

    ii. After coating the Au with a positive photoresist, the array of 100 15-, 60-, or 500-µm wide lines is patterned by UV exposure of the desired mask (b).

    iii. The exposed portions of the resist are dissolved in 0.35FNaOH; the newly exposed Au regions are wet etched for 5-10s in $0.2FI_2/1.2FKI$ to remove the 0.1-µm thick Au layer. The remaining resist is dissolved in acetone to reveal the array of Au line microelectrodes.

    B. Silk-screening metallo-organic inks

    i. Au or Pt metallo-organic ink [Englehard] is applied through a fine stainless steel screen of the desired image (i.e., pattern) to a 3mm Pyrex glass plate. The resultant noble metal film thickness is varied from 0.05-0.2µm by diluting the ink to obtain a lesser thickness.

    ii. The coated plate is fired at 675°C for 20min.

2. After an open-face Au or Pt linear array is formed, the submicrometer array is generated by covering one end of the surface up to an edge with a 1-mm thick glass plate. The cover plate is glued on with epoxy resin [Super Strength Araldite, Ciba-Geigy] and pressed against the array-glass substrate during the 24h cure of the epoxy; this produces a 10-15µm epoxy layer.

3. The front edge of the epoxied assembly is ground flush on a rotating diamond disk and polished by hand on successively

finer grades of wet sandpaper. A final polish is made with 1-μm alumina-water slurry and metal polish [Brasso] on paper.

4. Electrical contact is made at the opposing edge. The pattern for the array is designed so that the linear elements enlarge at the connection end; this permits easier electrical contact to an individual electrode element. Obvious care is taken to minimize scratching the thin metal layer at the point of contact. Single-electrode contact is made with an alligator clip modified so that the top jaw is a silver alloy tip to act as the connecting surface and the lower jaw is plastic. Electrical contact to the entire array or portions thereof is made with an aluminum shorting bar coated with conductive silicone rubber.

(a) Anderson and Sanderson (43) find that adhesion of their Cr/Au layers on quartz substrates is improved by annealing at 300-400°C for 1-2h after vapor deposition of the metals.

(b) Wrighton et. al., have designed 2-8 individually addressable line ultramicroelectrodes using photolithographic patterning of a positive photoresist coated on p-Si/SiO$_2$/Si$_3$N$_4$ substrates (40,46 and references therein). The patterned sample is then coated with a 6nm adhesive layer of Cr and a 120nm layer of Au. The line ultramicroelectrodes are developed by metal lift-off after exposure to warm acetone. The width of the Au line has been as narrow as 1.2μm with an insulating gap of 0.9μm. To improve the reproducibility of the electrochemical response of the array, Pt can be electroplated over the Au from a solution of 2mFK$_2$PtCl$_4$ in 0.1FK$_2$PO$_4$ (41,46,47,80). Controlling the amount of charge during the electrodeposition controls the amount of Pt electrodeposited and allows line widths, and thus, the spacing between lines to be varied symmetrically or asymmetrically.

Pt
Microlithographic patterning of 0.3μm thick Pt films: interdigitated array of 20 electrode pairs with a finger width of 3.5μm, separated by 2.5μm insulating gaps (44)

1. The borosilicate glass substrate is rinsed with a jet of distilled water, sonicated in 2-propanol for 15min, suspended over refluxing 2-propanol for 12h, and dried.

2. A 100-nm thick Ti layer is sublimed onto the substrate at 0.2nm/s from a resistively heated Ti coil placed 15cm below the substrate face.

3. A 200nm-thick Pt layer is vapor deposited at 0.2nm/s from a pre-sputtered Pt foil; a mirrored surface results.

4. The glass/TiO_2/Pt surface is coated with photoresist and photolithographically patterned commercially [Sawtek Inc., Orlando, FL].

5. The substrate is scribed, broken, sprayed with H_2O, and dried at 120°C for 30min.

6. The resist is Ar^+-ion etched followed by a wet Ti-etch (5min at 60°C in 100g H_2O:10g 30%H_2O_2:4g H_4EDTA:4g 30%NH_4OH); the remaining resist film is removed in a low-power oxygen plasma.

7. The array face is recoated with photoresist to protect the surface during mounting and scribing. The patterned substrate is rubber cemented on a glass block and cut into individual interdigitated array (IDA) regions with a diamond saw. Each patterned 5cm x 5cm glass substrate yields 192 IDA regions, each region containing 20 interdigitated electrode pairs.

8. Each IDA region is tested for electrical shorting; if shorting occurs, a 20-40V_{AC} bias often effectively melts the shorting Pt filaments.

9. An IDA is epoxied [Scotch 5-min epoxy] to cylindrical mounts [prepared from Buehler casting epoxy] that contain two Cu wires running the length of the mount. The protective photoresist is dissolved in methyl ethyl ketone or developer.

10. Indium solder is melted to connect each contact pad to one of the Cu leads; the Cu and In are covered with 5-min epoxy. All exposed epoxy and contact pad regions are meticulously masked with Apiezon vacuum wax dissolved in cyclohexane. This is usually done under a stereoscopic microscope to leave only the Pt fingers exposed.

11. The Pt surface of the IDA is cleaned by brief cycling from -0.12 to +1.4V(SSCE) in 1FH$_2$SO$_4$; the voltammogram is characteristic of bulk, polycrystalline Pt.

Irregular arrays of irregularly shaped microelectrodes

As discussed previously, extending the total electrode area while maintaining microelectrode characteristics is analytically and electrosynthetically useful and experimentally desirable since currents typical of macroscopic voltammetric electrodes are obtained and conventional instrumentation can be used. Fabricating regular arrays can be technically complex and exacting or merely tedious. Several workers have studied easily constructed composites of carbon and a polymeric insulator which yield arrays of nonuniformly sized, irregularly shaped microelectrodes. The two approaches to date either grind or cast small graphite particles with a polymer or perfuse porous carbon with an epoxy.

C

Perfusing insulating epoxy into porous C: irregular arrays of reticulated vitreous carbon (RVC) (50)

1. A 1-cm thick coring is punched from 45,60,80, or 100 pores/linear inch RVC [Normar Industries, Anaheim, CA] with a 6-mm i.d. glass tube. The plug is cleaned with an air jet to remove loose particles.

2. The RVC plug is lowered into an acrylic mold (6mm i.d.) filled with nonconductive epoxy [Buehler epoxide resin and hardener], so that a 2-mm space remains above the surface of the epoxy. After curing for 8h, the space is filled with conductive epoxy and a small nut is inserted for electrical contact.

3. After 8h, the carbon-epoxy plug is removed from the mold, centered in a 10-mm i.d. mold, and surrounded with nonconductive epoxy to cover the top surface of the contact nut.

4. After a final curing step, the electrode array is removed from the mold. The surface of the array is exposed by cutting 2-3mm off of the end with a diamond saw.

5. The array surface is mechanically polished with successively finer grades of diamond compound from 45μm to 0.25μm and finished with 0.05μm alumina.

6. For use, the array is screwed into an electrode shaft sheathed by the same nonconductive epoxy.

C
Mixing graphite particles with nonconductive epoxy: irregular composite array with 1μm sized C particles (53)

1. 45-50 weight percent of graphite powder ($\approx$1μm particle size) is thoroughly mixed with epoxy [Epotek 320] to make a thick paste {a}.

2. The composite paste is packed ~1cm deep into a 7mm glass tube and a Cu wire is inserted for electrical contact. The epoxy is cured for 2h at 80°C.

3. The surface is sanded flat and polished with 0.3 m alumina.

   {a} Weisshaar and Tallman (52) prepared graphite-Kel F mixtures which were hot-pressed under vacuum into cylindrical pellets for use as composite array electrodes. The graphite powder used [Ultracarbon Corp., Bay City, MI] sieved to less than 1μm, however, micrographs of the composite material show irregularly sized and shaped aggregates of graphite ranging from 10-200μm in width. Iontophoretic measurements by Engstrom and co-workers (53) also imply graphite aggregation in their graphite-epoxy composites.

4

Selected Reviews

James L. Anderson

Anderson and his colleagues have investigated the use of arrays of microelectrodes in systems in which a depolarizing fluid flows past the array. There is considerable interest in such systems as detectors, in particular as detectors in chromatographic analyses.

In one of their first studies, Anderson and Moldoveanu (1) carried out a numerical analysis of convective diffusion in a channel flow electrode. The objective of this work was to determine the current for a mass transfer limited electrochemical reaction which takes place at a plane electrode; the electrode is part of a rectangular channel. Steady state conditions and fully developed laminar flow apply. Anderson and Moldoveanu consider diffusion in the z-direction, normal to the electrode, and flow along the x-direction. The equation which governs the diffusion under conditions of fully developed laminar flow is

$$D\partial^2 c/\partial c^2 - v_x \partial c/\partial x = 0 \qquad [1]$$

in which the quantity v_x, the velocity of the fluid in the x-direction, is given by

$$v_x = (6U/bW_c)(z/b)(1 - z/b). \qquad [2]$$

In this expression, c is the concentration of the depolarizer, D is the associated diffusion coefficient, W_c is the width of the channel, and b is the distance between the working electrode and the opposite wall of the channel.

The electrode has finite length on one wall of the rectangular channel. The boundary conditions for a mass transfer limited reaction at the surface of the electrode therefore are

$x = 0$: $c = c^*$
$z = 0$: $c = 0$;   and $z = b$: $\partial c/\partial z = 0$.

For purposes of their subsequent analyses, Anderson and Moldoveanu convert to dimensionless units, the concentration being

$$g = c/c^*. \qquad [3]$$

The differential equation is converted to a form suitable for numerical analysis, namely, a finite difference representation. The principal numerical methods applied to the solution of this equation were the backward implicit method and the Crank-Nicholson implicit method. From the results of the analyses, the authors were able to evaluate an expression for the current:

$$I = W_e nFDc^* \Sigma_k g_{1,k}(\Delta x/\Delta z) \qquad [4]$$

in which $g_{1,k}$ is the dimensionless concentration in the cell ($z=1\Delta z$, $x=k\Delta x$).

The accuracy of their numerical analyses were assayed against theoretical (2) and semi-empirical (3) methods. The results gave good agreement with the semi-empirical results of

Weber and Purdy (3) and reasonably good agreement with the theoretical result (2).

Although the Anderson-Moldoveanu analysis relies on the division of the entire thickness of the channel into increments of uniform size, the authors argue that the results are indeed representative of real thin-layer channels. The same formulation applies under all conditions. There are two disadvantages which the authors point out: the computational method is inefficient in that a large fraction of the spatial elements experience no change in concentration and second, if the number of increments inside the diffusion layer becomes too small for accurate calculations, potential errors arise. These problems are pronounced for the simple explicit method of finite difference analysis. However, the simple backward implicit method, investigated by Anderson and Moldoveanu, was found to be remarkably insensitive to both of the problems.

Realizing their potential utility as chromatographic detectors, Anderson et al. (4), examined electrochemical microelectrode arrays as flow detectors. They compared the Kelgraf (a Kel-F and graphite composite) and glassy carbon electrodes. The Kelgraf electrode which behaves as an ultramicroelectrode array was found to be superior. It is particularly valuable as an oxidative flow detector.

Anderson et al. experimentally examined the flow response and other characteristics of the Kelgraf electrode. The mobile phase and solvent oxidation current can sometimes contribute a significant amount of noise in electrochemical detection systems. However, by controlling the active area of the microelectrode array of the Kelgraf electrode, it is possible to control these sources of interference. In addition, the capability of the Kelgraf electrode at extreme potentials was demonstrated by the use of the electrode to detect various carbamate pesticides by liquid chromatography using electrochemical detection (LCEC).

The hydrodynamic response of the electrodes was examined. By using the numerical analysis outlined in the preceding paragraphs, they found that the diffusion layer thickness for a solid electrode could be found at a distance X from the leading edge. However, the authors note that there has been little theoretical work to describe the flow response at microarray electrodes. With the use of an extension of the analysis, Moldoveanu and Anderson (5) examined several parallel strip arrays of electrodes. In addition, some of the arrays consisted of randomly distributed active sites, much in the spirit of the Kelgraf electrode itself. The randomly distributed arrays yielded responses which differed only a few percent from the predictions for an orderly strip array. As a result, Anderson et al. (4) concluded that it is possible to approximate the randomly distributed Kelgraf electrode reasonably well simply with an ordered array model.

Their experimental investigation of the hydrodynamic response yielded agreement with the $U^{1/3}$ law which follows from the theory (1). The background current response was found to be attributable to the oxidation of the solvent. The level of this response should vary directly with the percent of graphite in the electrode.

The Kelgraf electrode has a clearly different response for reactions which are controlled by mass transfer as compared to reactions which are controlled by electron transfer. This difference in response provides a means for the discrimination against the background current due to the oxidation of water or other surface-limited processes which take place at high applied potentials.

These observations were applied in a practical setting, as

mentioned, to detect the presence of carbamate pesticides in river water. A relatively high potential of +1.1V, which is needed to detect these types of compounds, strongly influences the detection limits and presents problems of potential background noise interference due to oxidative and other processes. For the Kelgraf electrode, however, excellent reproducibility was found for a given day's run. The electrode, nevertheless, was repolished every three to four days to ensure an optimum response.

The authors conclude that the Kelgraf electrode follows the theory for a microelectrode array. Moreover, they conclude that the electrode itself is well suited as a detector in flow detectors; it is better suited, they argue, than the solid electrode.

The theory of the amperometric response of a microarray electrode under conditions of convective diffusion was pursued further by numerical simulation (6). The actual simulation used the backward implicit finite difference (BIFD) method. The computational algorithm was extended in order to calculate the current and concentration profiles for the case of a simple mass transfer limited electron transfer reaction at an array electrode on one wall of a thin flow-through channel. Several geometries of the channel were used. In addition, steady state conditions were assumed together with the neglect of lateral and longitudinal diffusion relative to the direction of the flow. The same diffusion equation as used in reference 1 was used here. The numerical analysis also develops logically from that in reference 1.

Most interesting, perhaps, is the investigation of pseudo-randomness of the active elements of the array. These geometries were contrasted with regular arrays. A wide range of conditions was examined. The current responses of the various arrays were evaluated as functions of the number of active sites, the fractional blockage of the surface, and flow conditions, all relative to solid electrodes. The geometrical pattern of the array affects the response of the current; a regularly spaced array yields the maximum response for a given degree of partial blockage of the electrode and a constant number of microelectrodes.

The authors note that although no major generalizations may be drawn from their study, the results of several simulations consistently yielded slightly higher currents for regular microarrays with microelectrodes of constant widths and gaps with constant widths than for any of the other patterned geometries they considered.

Anderson, Ou, and Moldoveanu next considered the interesting case of a set of interdigitated microelectrodes with the alternate electrodes held at different potentials (8). As with their other studies, the electrode system occupies one wall of a flow channel. Because of the alternating sequence of array electrodes, it is now possible to consider both oxidation and reduction as the depolarizing fluid passes along the channel. Thus, two equations of the form of Equation [1] are required. The velocity of fluid flow is given still by Equation [2]. However, in addition to the geometric boundary conditions, there is a coupling of the fluxes through the equations

$$D_O(\partial c_O/\partial z) + D_R(\partial c_R/\partial z) = 0. \qquad [5]$$

Upon carrying out the numerical analyses for a variety of configurations, Anderson et al. concluded first that agreement with previous work for a single pair of electrodes under comparable conditions was very good. They find that a substantial improvement in the signal-to-noise ratio is predicted for the interdigitated electrode array relative to the single generator-detector pair of electrodes of equal overall area. This relative enhancement increases significantly with the number of generator-detector pairs present.

Finally, Fosdick and Anderson (8) undertook to find the optimum geometry for a microelectrode array flow detector when operated at a constant applied potential. The array, as with their earlier work, was located on one wall of the rectangular channel. Flow in the channel was laminar. They found that the optimum electrode response is obtained with a microelectrode array which has active sites of constant length separated by uniform gaps.

References

1. J. L. Anderson and S. Moldoveanu, J. Electroanal. Chem., 179 (1984) 107.
2. S. Moldoveanu and J. L. Anderson, J. Electroanal. Chem., 175 (1984) 67.
3. S. G. Weber and W. C. Purdy, Anal. Chem., 100 (1978) 531.
4. J. L. Anderson, K. K. Whiten, J. D. Brewster, T.-Y. Ou, and W. K. Nonidez, Anal. Chem., 57 (1985) 1366.
5. S. Moldoveanu and J. L. Anderson, J. Electroanal. Chem., 185 (1985) 239.
6. J. L. Anderson, T. Y. Ou, and S. Moldoveanu, J. Electroanal. Chem., 196 (1985) 213.
7. L. E. Fosdick and J. L Anderson, Anal. Chem., (submitted for publication, Jan. 1986).

Allen. J. Bard

Bard has made experimental and digital simulation investigations of ultramicroband (line) arrays. Digital simulation of collection, shielding, and feedback experiments with arrays were in good agreement with experimental results obtained by Wrighton and co-workers using 1-3μm-wide band electrodes spaced 1-3μm apart (1). Such arrays were also used for the study of homogeneous reactions coupled to the electrode reactions. Preparation of electrodes was described which was based on sputter deposition of thin platinum films on both sides of thin (2-12μm) plates of mica (2). Platinum band electrodes were prepared that ranged from 0.01 to 6μm thick, and 0.5 to 1.2cm long. An extensive set of experiments on model systems demonstrated the reliability of these devices for electroanalysis. Digital simulations of the results, including the effects of electrode shielding, were performed, and were found to be both consistent with established data, as well as suitable for the determination of second order homogeneous rate parameters. The reaction of ferricyanide ion with ascorbic acid (pH 2.5) and aminopyridine (pH 14) at 25°C was found to be 27±4 l-mol^{-1}-s^{-1} and 8±1x10^2 l-mol^{-1}-s^{-1}, respectively.

References

1. A.J. Bard, J.A. Crayston, G.P. Kittlesen, T. Varco Shea, M.S. Wrighton, Anal. Chem., 58 (1986) 2321.
2. T. Varco Shea and A.J. Bard, Anal. Chem., submitted.
3. H.-Y. Liu, F.-R. F. Fan, and A.J. Bard, J. Am. Chem. Soc., Comm. Ed., 108 (1986) 3838.
4. C.W. Lin, F.-R. F. Fan and A.J. Bard, J. Electrochem. Soc., in press.

Alan M. Bond

Alan M. Bond has been involved in several areas of ultramicroelectrode electrochemical research (see also the reviews on Bixler, Pons, Fleischmann, and Thormann). In addition to those works, he has been involved in the following projects.

One study was directed at alternating and direct current measurement methods for undertaking studies of adsorption processes in the absence of added electrolyte (1). In that work, the effect of decreasing the added electrolyte (NaF) concentration from 0.9$\underline{M}$ to zero on the adsorption of courmarin (from its saturated solution at 15°C) at dropping and stationary mercury electrodes was studied. The workers used alternating current voltammetry, polarography and cyclic voltammetry at conventional sized electrodes and at ultramicroelectrodes. In all cases the adsorbed monolayer of courmarin at the electrode was found to form three surface phases in the potential range studied and the data are consistent with that predicted by theory. Reliable quantitative data was obtained for solutions containing greater than $10^{-2}\underline{M}$ concentrations of sodium fluoride at conventional sized electrodes. At lower or zero added electrolyte concentrations, only qualitative data could be obtained by DC methods at conventional electrodes. Furthermore, essentially no useful AC response was observed because of the high level of uncompensated resistance present even when a three electrode potentiostat was used with positive feedback circuitry. A much improved cyclic voltammetric response was obtained at low or zero added electrolyte concentrations when a hemispherical mercury ultramicroelectrode with a 5μm radius was used. The ultramicroelectrode was prepared by deposition onto platinum. Although the alternating current voltammetric response using the ultramicroelectrode was vastly superior to that from a conventional sized electrode in the absence of added electrolyte, the task of accurately measuring pF capacitances in the presence of MΩ resistances by alternating current techniques is still difficult.

Bond has investigated ultramicroelectrode electrochemistry in aromatic hydrocarbons (benzene, toluene and xylene) which represent a class of solvents of considerable interest in electrochemical research and technology. Unfortunately, voltammetry in these high resistance aromatic hydrocarbons is complicated by ohmic iR drop and other factors when using conventional sized electrodes or even ultramicroelectrodes (radius > 1μm). The use of hexylammonium perchlorate as an electrolyte and ultramicroelectrodes with radii less than 1μm enables the reversible diffusion controlled theoretical response to be observed for the oxidation of ferrocene in benzene, toluene and xylene under steady state conditions, at low scan rates (<100mV·s^{-1}). (See also the review of Parker herein.) Ambient temperatures can be used with benzene and toluene. Slightly elevated temperatures (45°C) are required to increase the solubility of the electrolyte in xylene. With larger radii ultramicroelectrodes or at faster scan rates, a peak shaped rather than sigmoidal shaped reduction response is commonly observed in benzene and toluene on the reverse scan of cyclic voltammograms. Tentatively, this nonsteady state response has been attributed to nucleation and precipitation of insoluble ferrocenium perchlorate onto the electrode surface. Under conditions of completely spherical diffusion, which is essentially the case at ultramicroelectrodes, these precipitation reactions are believed to occur too far away from the electrode surface to influence the voltammetric response. The importance of electrode size, scan rate and electrolyte concentration as factors in obtaining distortion free voltammetric data in aromatic hydrocarbons will be described in detail in the published report.

Bond, Thormann, and van den Bosch (3) described the behavior of gold and platinum linear voltammetric microelectrode arrays constructed by lithographic technology. The workers claimed that the individual elements are geometrically equivalent to circular disk ultramicroelectrodes of radii 0.6-5μm. Voltammetric measurements were made at the individual sensing elements, at part of the array, and at the total linear ensemble. The response of the array was close, but not exactly equal, to the sum of the individual sensing elements. A steady state voltammogram was observed with current in the microampere range (for the oxidation of 1m$\underline{M}$ ferrocene in acetonitrile) for the larger arrays. All constructed ensembles exhibited microelectrode characteristics that permitted measurements to be made in the absence of deliberately added electrolyte. Additionally, transient techniques such as ac and pulse voltammetry can be applied with millisecond pulse widths in systems with supporting electrolyte and with high resistance. Commercially available instrumentation can readily be used with the larger arrays and was described in detail in that work.

Bond and Thormann (4) have also investigated the reversible one-electron oxidation of ferrocene in acetonitrile at arrays of microdisk electrodes, linear microelectrodes and very thin ring electrodes prepared with a uniform glass seal or a layer of epoxy resin between the conductor and the glass support in the ms and sub-ms time domains by a range of transient electrochemical techniques and chronocoulometry. The data reported were compared with those from a conventional sized disk electrode (0.8mm radius) and the comparison illustrated the significant advantages of using ultramicroelectrodes with transient voltammetric techniques provided that the electrodes were correctly constructed. The responses of the transient

electrochemical techniques of differential pulse, square wave
and ac voltammetry, as well as chronocoulometry were strongly
influenced by irregularities originating in the construction of
inlaid microelectrodes of microscopic dimensions of less than
1μm. The authors pointed out that imperfectly prepared
electrodes with cracks in the glass seal or with thick sealing
layers of epoxy resin exhibited undesirable background currents
which prevented their use for measurements at short time scales.

Bixler, Bond (5), and others demonstrated the practicality
of using ultramicroelectrodes for detection of small amounts of
analyte. They pointed out that ultramicroelectrodes should
provide greater analytical sensitivity than electrodes of
conventional size. However, the detection of micromolar or
lower concentrations with microdisk electrodes requires
measurement of femtoamp currents, which is outside the range of
most commercially available instrumentation. The novel use of a
picoammeter or femtoammeter as a current amplifier permits
commercial instrumentation to be used with ultramicrodisk
electrodes. Such instrumentation incorporating a picoammeter or
femtoammeter is limited by the relatively slow rise time; this
places restrictions on scan rates in all voltammetric techniques
and on pulse widths in transient techniques such as differential
pulse and square wave voltammetry. Because of the small
currents, the ohmic (iR) drop is very small and polarization of
the reference electrode is unimportant; thus a two-electrode
format without a potentiostat can be used. Consequently, a
microprocessor-based function generator and data storage system,
in conjunction with a pico- or femtoammeter, is satisfactory in
providing inexpensive, versatile and very sensitive instrumentation for voltammetric detection with microdisk electrodes.
Convenient methods for fabricating platinum, gold and carbon
microdisk electrodes for use in stationary and flowing solution

configurations were also presented in that work.

Bond et al. (6) investigated the theory of cyclic voltammetry at linear gold microelectrodes of thicknesses less than 1μm for electrode processes where the electron transfer step is rate determining. Digital simulation (expanding-grid method) was used with cylindrical diffusion in the theoretical calculations. Results of the theory in the limit of the reversible response were in agreement with previously published theory and with experimental data obtained in the work. Unlike ultramicrodisk electrodes, departures form the steady state sigmoidal shaped curves were predicted and observed experimentally (provided considerable attention was given to electrode design). The very thin electrodes achieved the property of having almost purely cylindrical diffusion and therefore are suitable for use in high resistance solutions and for measurements of very fast rates of electron transfer. The large total area, provided by the dimension of length, means that measured currents can be sufficiently large so that transient voltammetric methods may be implemented with conventional instrumentation.

References

1. A.M. Bond and F.G. Thomas, private communication and to be published.
2. A.M. Bond and T.F. Mann, private communication and to be published.
3. W. Thormann, P. van den Bosch and A.M. Bond, Anal. Chem., 57 (1985) 2764.
4. W. Thormann and A.M. Bond, J. Electroanal. Chem., 218 (1987) 187.
5. J.W. Bixler, A.M. Bond, P.A. Lay, W. Thormann, P. van den Bosch, M. Fleischmann and S. Pons, Analytica Chimica Acta, 187 (1986) 67.
6. A.M. Bond, T.L.E. Henderson and W. Thormann, J. Phys. Chem., 90 (1986) 2911.

Martin Fleischmann

Recently, Fleischmann and co-workers have demonstrated that electrochemical measurements in solutions containing no purposely added supporting electrolyte is feasible at ultramicroelectrodes (1). A two-electrode configuration was used to eliminate the noise problems that are associated with three-electrode potentiostats. The shape of the voltammetric wave shows some distortion due to ohmic losses. The effect has been treated theoretically (2). In principle, these measurements could lead to the study of the theoretically interesting one ion problem, but traces of impurity ions and water probably contribute to a dilute electrolyte pool. The authors point out that the use of ultramicroelectrodes allows electrochemical studies of organic reactions to be made in highly resistive organic solvents, measurement of diffusion coefficients at zero ionic strength, and electrochemical detection in chromatographic applications.

Fleischmann et al. demonstrated that at platinum ultramicroelectrodes with radii less than 1 micron it is possible to observe voltammetry at temperatures down to the freezing point of the solution (eutectic mixtures) (3). Results were obtained with and without purposely added supporting electrolyte. At lower temperatures, a glass is formed, and voltammetry could still be performed. Such measurements are feasible due to the low ohmic losses associated with ultramicroelectrodes. Data were obtained for the oxidation of ferrocene in acetonitrile at temperatures from 25°C down to -78°C. Additionally, measurements were made down to the eutectic temperature in acetone and dichloromethane. It was pointed out that such measurements allow the study of electron transfer reactions without the complications due to the following homogeneous chemical reactions following the charge transfer reaction.

Further work was concerned with a new technique for the determination of the kinetics of fast first order, or pseudo first order, homogeneous chemical reactions that are coupled to electron transfer reactions (4). Steady state voltammetric plots were recorded at platinum ultramicroelectrodes and an analysis was developed for the limiting current region that leads to the determination of the kinetic parameters. The ce, ec', ferrocyanide reduction (in the presence of amidopyrene), and hydrogen evolution reactions were studied. The rate parameters obtained agreed well with those obtained by other methods.

The second part of this study was directed at other coupled reaction mechanisms (5). The studies were made on the oxidation of anthracene and of hexamethylbenzene in dry acetonitrile solutions. These oxidations were compared with those performed by conventional methods, and were found to be ece and displ reactions respectively. The analysis leads to determination of rate parameters from the gradients of simple straight line "working curves" of the inverse of the limiting current against the electrode radius. The parameters obtained were in close agreement with those obtained by conventional means. It was shown also that the problem of nonuniform accessibility across the surface of small disk electrodes leads to complications in the determination of kinetic parameters for second order reactions. Fleischmann and others published a forward looking review on the construction and behavior of ultramicroelectrodes (6). Methods of construction and the behavior of microdisk, microsphere, and microring electrodes were described.

For other recent work, the reader should consult the reviews herein on Pons and Bond (7-12).

References

1. A.M. Bond, M. Fleischmann, and J. Robinson, J. Electroanal. Chem. 168 (1984) 299.
2. A.M. Bond, M. Fleischmann, and J. Robinson, J. Electroanal. Chem., 172 (1984) 11.
3. A.M. Bond, M. Fleischmann and J. Robinson, J. Electroanal. Chem., 180 (1984) 257.
4. M. Fleischmann, F. Lasserre, J. Robinson, and D. Swan, J. Electroanal. Chem., 177 (1984) 97.
5. M. Fleischmann, F. Lasserre and J. Robinson, J. Electroanal. Chem., 177 (1984) 815.
6. A. M. Bond, M. Fleischmann, S. B. Khoo, S. Pons, and J. Robinson, Ind. J. Tech., 24 (1986) 492.
7. M. Fleischmann, S. Bandyopadhyay and S. Pons, J. Phys. Chem., 89 (1985) 5537.
8. M. Fleischmann, J. Ghoroghchian, and S. Pons, J. Phys. Chem., 89 (1985) 5530.
9. J. Cassidy, S.B. Khoo, S. Pons, and M. Fleischmann, J. Phys. Chem. 89 (1985) 3933.
10. T. Dibble, S. Bandyopadhyay, J. Ghoroghchian, J. J. Smith, F. Sarfarazi, M. Fleischmann, and S. Pons, J. Phys. Chem., 90 (1986) 5275.
11. A. Russell, K. Repka, T. Dibble, J. Ghoroghchian, J.J. Smith, M. Fleischmann, C. Pitt and S. Pons, Anal. Chem., 58 (1986) 2961.
12. M. Fleischmann and S. Pons, J. Electroanal. Chem., 222 (1987) 107.

Michael Grätzel

Michael Grätzel and his group have been involved with charge transfer reactions at colloidal semiconductor particles. This type of investigation may be compared with charge transfer reactions driven at dispersed metal particles that are suspended in a highly resistive solution. (See other works in this document.) Driving of reactions has been traditionally accomplished by laser excitation of a chromaphore which subsequently injects charge into the conduction band of the colloidal semiconductor. In one investigation (1), Grätzel reported the first picosecond time-resolved kinetic study of a photosensitized charge injection from an excited state dye (eosin Y) into the conduction band of 12nm TiO_2 particles after excitation of the dye with a laser. Grätzel thus *directly* observed charge injection, with a rate constant of $(9.5\pm1.4)\times10^8 s^{-1}$.

In other work (2), Grätzel and co-workers have investigated the dynamics of the charge transfer reaction in the reduction of cobaltocenium-dicarboxylate in colloidal suspensions of TiO_2. In this work, the semiconductor was excited with a laser, and the kinetics of charge injection to the electrophile were determined by measurement of the loss of the blue color of the electron in the semiconductor, and formation of the reduced chromophore. At pH 10, the electron spectrum had a maximum at 780nm, with an extinction coefficient of $800 M^{-1} cm^{-1}$. Electron transfer from the conduction band to other chromophores (viologens) was found to be greatly enhanced by the dicarboxylate as the acid strength of the solution was increased. Adsorbed dicarboxylate at the semiconductor-solution interface was found to be responsible for this catalytic effect.

In another study (3) very high sensitizations of anatase TiO_2 particles and polycrystalline electrodes could be attained using tris(2,2'-bipyridyl-4,4'-dicarboxylate)ruthenium(II) dichloride as the sensitizing agent. A quantum yield of 60±10% for injection could be obtained. 44% of the incident photon energy could be converted to current, which we believe is the highest attained for any system as of this date.

Grätzel has also studied the kinetics of charge carrier trapping and electron-hole recombination reactions in TiO_2 colloidal particles (4). The temporal study of the absorption spectrum of the trapped electron formed from laser excitation was found to lie within the laser pulse profile, while the hole trapping was much slower ($\approx$250ns). The recombination rate was determined to occur as a second order process with k = (3.2±1.4) x $10^{-11} cm^3 s^{-1}$ at high densities of electrons and holes. Intraparticle recombination at low reactant density follows first order kinetics where a pair has a characteristic lifetime of (30±15)ns. It is likely that valence band hole trapping at low densities (by reaction with surface hydroxyls) competes with the recombination process. These results were used to explain earlier observations, where the lifetime of the charge transfer from TiO_2 to methylviologen was very much slower in acid solutions; hole trapping by surface hydroxyls which leads to products slow to react with electrons explains this result.

References

1. J. Moser, M. Grätzel, D.K. Sharma, and N. Serpone, Helv. Chim. Acta, 68 (1985) 1686.
2. U. Kolle, J. Moser and M. Grätzel, Inorg. Chem., 24 (1985) 2253.
3. J. Desilvestro, M. Grätzel, L. Kavan and J. Moser, J. Am. Chem. Soc., 107 (1985) 2988.
4. G. Rothenberger, J. Moser, M. Grätzel, N. Serpone and D.K. Sharma, J. Am. Chem. Soc., 107 (1985) 8059.

Jonathon O. Howell

J. Howell et al. (1) have evaluated methods for the elimination of residual current in fast scan (>300V-s^{-1}) voltammetry performed at microvoltammetric electrodes. It was found that only a moderate improvement was gained in the ratio of the faradaic current to the residual current by using staircase voltammetry. By digitally subtracting residual current (determined from the response in supporting electrolyte solution) from the voltammogram obtained in the presence of the electroactive species, improved results were obtained. The subtraction process is expedited by performing the entire experiment in a flow injection apparatus in such a way that the electrode remains in solution while the two voltammograms are obtained. Undistorted voltammograms were obtained for the reduction of anthracene (~2m$\underline{M}$) at scan rates up to 10,000V-s^{-1}. Essentially undistorted voltammograms were obtained for the reduction of Tl$^+$ at low concentrations (20$\mu\underline{M}$) in aqueous solution at a mercury microvoltammetric electrode. The dynamic range was improved by analog offset of the residual current. It was found that this technique is sensitivity limited by electronic noise.

Reference

1. J.O. Howell, W.G. Kuhr, R.E. Ensman and R.M. Wightman, J. Electroanal. Chem., 209 (1986) 77.

Richard McCreery

Richard McCreery has investigated the feasibility of spectroelectrochemical techniques at ultramicroelectrodes. Robinson and McCreery (1) monitored scattered light from 12μm diameter carbon fibers and observed transient absorbances for formation of o-dianisidine dication, and measured the absorbance spectrum for methylviologen cation radical. It was determined that the absorbance time response for the microcylinder is the same as that for reflection from a plane.

McCreery and co-workers extended the work on microcylinders to gold and platinum devices (2). In this work, the measured absorbances were compared to responses predicted from well known models from heat transfer literature. It was found that the temporal absorbance response was in good agreement up to about 0.3s. Deviations above this limit could be reduced by lowering the laser power. It was demonstrated that reaction kinetics using this technique could be made at the microsecond level. In later work McCreery et al. (3) reported that the technique is established at < 40ns.

Schuette and McCreery investigated square wave voltammetry at platinum microdisk electrodes. They predicted and observed that the average ac current was linear with modulation frequency and concentration. Ferrocene was shown to have a detection limit of 2×10^{-7} M.

References

1. R.S. Robinson and R.L. McCreery, Anal. Chem., 53 (1981) 997.
2. R.S. Robinson, C.W. McCurdy, and R.L. McCreery, Anal. Chem., 54 (1982) 2356.
3. R.S. Robinson and R.L. McCreery, J. Electroanal. Chem., 182 (1985) 61.
4. S.A. Schuette and R.L. McCreery, J. Electroanal. Chem., 191 (1985) 329.

Marcin Majda

Marcin Majda has developed a novel, structurally microporous electrode coating template of aluminum oxide which has important implications in the area of heterogeneous electrocatalysis (1). Porous aluminum oxide templates are produced by electrooxidation of aluminum in phosphoric acid, where the steady state growth of the oxide film results in an almost perfectly regular array of densely packed, cylindrical pores penetrating the oxide layer. When the growth of the film reaches a desired thickness, the oxide layer is dislodged by amalgamation and treated chemically to dissolve the impervious barrier layer. Subsequently, vapor deposition of gold onto one side of the microporous aluminum oxide template and mounting at the tip of a glass tubing produces an array of gold microelectrodes coated with a porous Al_2O_3 film (2).

Immobilization of catalysts involves chemical derivitization of the inner surfaces of the porous oxide template. This type of electrode provides a large surface area for catalyst immobilization and retains substantial permeability of the electrode films (approximately 40 to 60%) due to the internal structure of the alumina templates.

Majda then investigated a number of reagent immobilization schemes and catalytic reactions. In one report, the porous aluminum oxide films were derivatized by adsorption of poly(4-vinylpyridine) along the walls of cylindrical oxide channels (2). Ferricyanide ions were bound electrostatically in acidic media to pyridinium sites in the polymer. The electron transport process, which accounts for the total electroactivity of all immobilized centers, involves two dimensional charge propagation along the oxide channels. More detailed chronoamperometric and rotating disk electrode studies showed that the electron transport relies also on ferricyanide diffusion in the aqueous phase and a fast electron exchange between electroactive species in the liquid phase and the surface, polymeric phase.

A different reagent immobilization scheme involved spontaneous self-assembly of amphiphilic molecules at the walls (3). In one example, an organized bilayer microstructure was produced by two consecutive self-assembly steps. The first was deposition of a monolayer of octadecyltrichlorosilane (OTS) onto inner aluminum oxide surfaces, followed by the second one which involved the octadecyl derivative of methylviologen ($C_{18}MV^{2+}$). This unique approach, combining microarray techniques with surfactant self-assembly methodology, resulted in bilayer assemblies with a perpendicular orientation with respect to the electrode surface thereby allowing the first direct electrochemical measurements of the lateral electron transport in organized monolayer systems (3,4).

Further studies were concerned with several different amphiphiles and addressed the question of the mechanism of electron transport and the dependence of its kinetics on a number of structural and environmental parameters (4).

References

1. C.J. Miller and M. Majda, J. Am. Chem. Soc., 107, (1985) 1419.
2. C.J. Miller and M. Majda, J. Electroanal. Chem., 207, (1986) 49.
3. C.J. Miller and M. Majda, J. Am. Chem. Soc., 108, (1986) 3118.
4. C.J. Miller, C.A. Widrig, D. Charych and M. Majda, J. Phys. Chem., submitted.

Barry Miller

Barry Miller and co-workers have investigated methods of generation of, and properties of, localized semiconductor ultramicroelectrodes (1). The electrodes were generated at chosen positions on semiconductor surfaces by a focussed laser. Light of the proper energy will cause formation of electron hole pairs in the semiconductor surface; their separation in an electric field leads to photoreactions initiated by the minority carrier if the bandgap energy is such that its edges encompass the appropriate solution energy levels of the redox couple. "Disk" ultramicroelectrode geometries on a semiconductor surface can thus be readily generated by focussing a laser beam on its surface. The workers used n-InP and N-GaAs semiconductor substrates. The workers evaluated the technique by using the semiconductor as the central disk, and gold as the ring collector in rotating ring-disk experiments. They showed that current could indeed by spatially controlled on the surface of the semiconductor by the procedure. Confirmation was made by measurements of transit time of species between the light generated disk and the ring, and by the collection efficiency. By both methods, Miller showed that the current stimulated by the focussed beam originated in the area fixed by the diameter corresponding to that determined through beam profile and disk edge crossing effects. The properties of the laser scanned beam were extended to a three dimensional representation of the entire semiconductor surface; the plot was presented as the ring current as a function of the x-y spatial representation of the disk surface. Two major points were demonstrated in this work. First, it is apparent that this technique is ideally suited for many types of kinetic measurements since it is possible to vary two time related properties: the hydrodynamic parameter (rotation speed of the rotating ring-disk electrode system) and the location and size of the photogenerated electrode (the size of which determines mass transport rate, and the location of which will also determine transit time. The authors mention applications to electron transfer coupled following chemical reactions and adsorption studies via the transit time technique. In addition there is an application towards measurement of corrosion currents or redox transfer efficiency when using the photocurrent mapping technique mentioned above.

Second is the feasibility of generating ultramicroelectrodes of specific shapes for controlling mass transport. Parasitic photocorrosion reactions will limit the ultimate speed and utility of these ultramicroelectrodes on exposed semiconductor surfaces. Overlayers to protect the surface, or the use of buried channel or junction devices may overcome this problem.

Reference

1. B. Miller and J.M. Rosamilia, J. Electrochem. Soc., $\underline{132}$ (1985) 2621.

Royce Murray

In a series of papers, (1-9) Murray and co-workers have investigated the simultaneous use of microelectrodes with thin polymer films. An objective of this work has been to take advantage of the fact that microelectrodes require substantially less supporting electrolyte than is the case for typical macroscopic devices. Thus, it was possible to observe the electrochemical activity of solutes in a variety of solvents without some of the complications of additional ionic interactions. In particular, microelectrodes can be used to avoid adverse potential gradient effects which are associated with extremely dilute solutions. The research being carried out suggested the possible development of devices which, while dependent upon chemical transformations and processes, would act in a manner similar to a number of solid state semiconductor devices: e.g., rectifiers, optical sensors, etc. There is thus an evolution toward a macromolecular electronics; the state of this particular application has been summarized by Chidsey and Murray (6).

In a study of the permeability of solutes through electroactive polymer films, 10μm-diameter electrodes were fabricated from Wollaston wire (1). The workers determined the permeability of 15-100nm-thick films of polycationic poly(tris(4-methyl-4'-vinyl-2,2'-bipyridine)ruthenium to ferrocene, ferrocenium, tetracyanoquinodimethane (TCNQ) in 0.1-10^{-5}$\underline{M}$ Et$_4$NClO$_4$ electrolyte in acetonitrile. Twin electrode, thin layer cells were used.

The authors also reported that the permeabilities vary with electrolyte concentration. This fact allows one to separate the contributions due to partition and diffusion of the electroactive permeant into and within the polymer. It was found that permeability is independent of the concentration of neutrals, ferrocene and TCNQ; permeability decreases for the cationic ferrocenium species and increases for anionic $TCNQ^-$. The change in permeability with the concentration of electrolyte depends principally on the partition coefficient of the permeant into the polymer.

The next step in their work (2), was to investigate the use of coated microelectrodes to study the kinetics of the reaction

$$Pt/poly\text{-}[M]^{3+} + [NL_3]^{2+} \xrightarrow{k_{12}} Pt/poly\text{-}[M]^{2+} + [NL_3]^{3+}$$

in which poly-$[M]^{2+}$ is an electropolymerized film of a metal (Fe, Ru, Os) polypyridyl complex on a Pt electrode. $[NL_3]^{2+}$ is a different, monomeric metal polypyridyl complex which was dissolved in the solution in contact with the polymer-coated electrode. The objectives here were to verify rates of reaction previously observed at rotated disk electrodes and to explore the capabilities and limitations of the micro and twin electrode geometries relative to the rotated disks.

The M(III) states of poly-$[Ru(vbpy)_3]^{2+}$ and $[Os(bpy)_2(vpy)_2]^{2+}$ films on Pt disk microelectrodes were used to mediate the oxidations of the other metal polypyridyl complexes. Steady state mediation currents in twin electrode thin layer cells could be observed even when the formal potential of NL_3^{2+} was as much as 570mV more positive than that of the mediating polymer film. The rate constants measured at the polymer-solution interface agreed with the rotated disk data, where available, and with the established Gibbs energy/rate correlation for others.

The authors conclude that the twin electrode thin layer cells are suitable to use in the study of very slow mediated reaction rates. The ultimate limit of the experiment occurs in the sensitivity to film imperfections in evaluating solute permeation and in the lack of data for the partition coefficients for these systems. Closing on a pessimistic note, Murray, et al. stated that getting round these difficulties for the reaction scheme shown above is not going to be easy, perhaps not even likely.

An interesting application of both microelectrode techniques and polymer film modifications was the measurement of the electron diffusion coefficients (3). In this study, the polymer films contained Prussian Blue as the electroactive species. The diffusion coefficient was actually measured with the use of an interdigitated array over which the polymer and electroactive agent were coated.

The charge required to reduce Prussian Blue was determined by integrating the cyclic voltammetric [Fe(III/II) ⟶ Fe(II/II)] peak which is obtained by scanning the potentials of both terminals of the interdigitated array simultaneously from 0.5V to -0.2V vs SSCE. The steady state current generated by scanning the potential of one IDA terminal from 0.5 to -0.2V while maintaining the other terminal at 0.5V was also measured. The limiting current was used together with the charge to determine D_e from the equation

$$D_e = i_{lim} dpN/Q(N-1)$$

in which Q is the charge collected, d is the thickness of the film of Prussian Blue, p is the center-to-center electrode spacing, and N is the number of finger electrodes.

The assumption that the IDA current flux was approximated as flowing through a film of height t and of width d, sandwiched between two parallel plate electrodes led to the prediction that D_e should be independent of t. This was verified. Thus, the parallel plate theory adequately and accurately accounts for the behavior of the electroactive material.

The aspiration to carry out electrochemistry in solvents of vanishingly small polarity seems to be within grasp through the use of microelectrodes. Geng, et al. (5) reported the study of electrochemical reactions of both solutes and polymer films in toluene and heptane. As supporting electrolytes, dinonylnaphthalenesulfonic acid/trioctylphosphine oxide and tetrahexylammonium perchlorate were used. Silver wire was used as a pseudoreference electrode; it was estimated as 0.29V positive of the SSCE reference electrode potential in acetonitrile.

The resistivity problem in heptane, for example, proved to be a much greater problem to overcome than did the resistance for ultradilute electrolyte in acetonitrile. The authors note that in heptane, TOPO or $DNNSO_3H$ are not conductive alone; no microelectrode currents are obtained. The mixture of the two species as an electrolyte, they suggest, acts as a charge carrier through acid/base reactions. Thus, ionic current in solution may exist as a result either of migration of the separated ions or by proton hopping between TOPO and $TOPOH^+$.

The authors were able to observe microelectrode voltammetry of 1m$\underline{M}$ ferrocene. There was evidence of film formation in the process of oxidation of this species.

The oxidations of other species, tetraphenylporphyrin and copper tetraphenylporphyrin have been investigated in toluene (7). Tetrahexylammonium perchlorate was used as the electrolyte. Enhanced axial coordination and acid-base chemistry was found in the oxidative electrochemistry of tetraphenylporphyrin and copper(II) tetraphenylporphyrin. The objective of this work was to try to understand why the porphyrins are oxidized in a single two-electron wave when sequential single electron steps are more typical. The single two-electron wave was resolved into two one-electron waves in higher concentrations of Hx$_4$NClO$_4$. Once again, this result demonstrates the strength of ultramicroelectrodes in determining mechanisms of electron transfer in highly resistive environments (see Chapter One).

Finally, in recent work, Murray and his colleagues (8,9) have been probing solid solutions with their microelectrodes. In particular, they have carried out voltammetry of electroactive solutes in polyethylene oxide polymer films which have been deposited on microelectrodes. It was possible to study various diffusion and reactive processes directly in these systems.

References

1. A. G. Ewing, B. J. Feldman, and R. W. Murray, J. Phys. Chem., 89 (1984) 1263.
2. B. J. Feldman, A. G. Ewing, and R. W. Murray, J. Electroanal. Chem., 194 (1985) 63.
3. B. J. Feldman and R. W. Murray, Anal. Chem., 58 (1986) 2844.
4. C. E. Chidsey, B. J. Feldman, C. Lundgren and R. W. Murray, Anal. Chem., 58 (1986) 601.
5. L. Geng, A. G. Ewing, J. C. Jernigan and R. W. Murray, Anal. Chem., 54 (1986) 852.
6. C. E. D. Chidsey and R. W. Murray, Science, 231 (1986) 25.
7. L. Geng and R. W. Murray, Inorg. Chem., 25 (1986) 3115.
8. R. A. Reed, L. Geng and R. W. Murray, J. Electroanal. Chem., 208 (1986) 185.
9. L. Geng, R. A. Reed, M. Longmire and R. W. Murray, J. Phys. Chem., (in press).

Janet Osteryoung

Janet Osteryoung and co-workers began work on ultramicroelectrode electroanalytical chemistry in the early 1980's. The first devices (1) they investigated were arrays of microdisk electrodes formed from photoresist covered glassy carbon electrodes. These devices were tested in hexacyanoferrate systems, and chronoamperometric responses were found to agree well with predictions that the authors had made previously (2) and with digital simulations of the disk geometry.

In addition, refinements to the theoretical bases and methods of preparation of ultramicroelectrodes were reported (3). In that work, platinum and gold disk electrodes were prepared by imbedding fine wires in epoxy resin in glass tubes. Chronoamperometric responses of these ultramicroelectrodes were found to follow the predictions made by Osteryoung and co-workers for a microdisk geometry, as opposed to the hemispherical approximation made by earlier workers.

Osteryoung et al. (4) investigated capillary shielding and spherical diffusion effects at small mercury drops for reverse pulse voltammetric experiments. Results were presented for the solution of the double potential step experiment at spherical electrodes. Aoki and Osteryoung (5) have presented a correction to their microdisk chronoamperometric theory. The theoretical results were extended to include the electroanalytical techniques of chronopotentiometry (6) and linear sweep voltammetry (7). They compared their theory to results obtained at electrodes $25\mu m$ or larger, i.e., in a size region where non-steady state diffusion was still important. Excellent agreement resulted.

Osteryoung and co-workers (8) presented an interesting method of producing ultramicroelectrode arrays from reticulated carbon cross sections by back-filling the spaces with epoxy resin. Details are given in the chapter on fabrication. The workers observed enhanced current due to edge effects.

In other work, Osteryoung and co-workers demonstrated that the form of square wave voltammograms were virtually independent of the geometry of the ultramicroelectrode (9).

Preparation of mercury coated carbon fiber electrodes (10) and mercury spheres on iridium microdisks (11) were reported, and are discussed in the fabrication chapter.

Hepel and Osteryoung (12,13) reported impedance relation results for chromium and gold ultramicroelectrode arrays. These electrodes (also described in the fabrication chapter) were 0.375μm in radius, and were prepared by electron beam lithographic methods. The electrodes numbered over $10^6 cm^{-2}$. At these very small electrodes, the impedance data demonstrated that the diffusion layers from adjacent electrodes did not interact in the frequency range studied.

Osteryoung, Aoki, and co-workers (14) derived an equation for the square wave voltammetric response of an ultramicroelectrode disk electrode. The peak shape and position was shown to be invariant to $D\tau/r^2$, where D is the diffusion coefficient, τ the square wave period, and r the radius of the disk, whereas the peak current intensity increases with increasing value of this parameter. The predicted response was obtained for the aqueous hexacyanoferrate system at a platinum electrode.

The Osteryoung effort also includes several works submitted and in press. These are concerned with linear scan voltammetry and chronoamperometry at small mercury film electrodes, determination of kinetic parameters from steady state microdisk voltammograms, and electrodeposition and anodic stripping of silver on single carbon fibers.

References

1. K. Aoki and J. Osteryoung, J. Electroanal. Chem., 125 (1981) 315.
2. K. Aoki and J. Osteryoung, J. Electroanal. Chem., 110 (1980) 19.
3. T. Hepel and J. Osteryoung, J. Phys. Chem., 86 (1982) 1406.
4. T.R. Brumleve and J. Osteryoung, J. Phys. Chem., 86 (1982) 1794.
5. K. Aoki and J. Osteryoung, J. Electroanal. Chem., 160 (1984) 335.
6. K. Aoki, K. Akimoto, K. Tokuda, H. Matsuda, and J. Osteryoung, J. Electroanal. Chem., 182 (1985) 281.
7. Ibid, 219.
8. N. Sleszynski, J. Osteryoung, and M. Carter, Anal. Chem., 56 (1984) 130.
9. J.J. O'Dea, M. Wojciechowski, J. Osteryoung, and K. Aoki, Anal. Chem., 57 (1985) 954.
10. J. Golas and J. Osteryoung, Anal. Chim. Acta, 181 (1986) 211.
11. J. Golas, Z. Galus and J. Osteryoung, Anal. Chem., 59 (1987) 386.
12. T. Hepel and J. Osteryoung, J. Electrochem. Soc., 133 (1986) 752.
13. Ibid, p. 757.
14. D. Whelan, J.J. O'Dea, J. Osteryoung and K. Aoki, J. Electroanal. Chem., 202 (1986) 23.

Vernon Parker

Vernon Parker was one of the first researchers (1977) to actively pursue organic reaction mechanisms in highly resistive media. The combined use of microelectrodes (large by today's standards) and reduced specific resistivity allowed some of the first electrochemical measurements to be made in benzene and chlorobenzene (1). High speed cyclic voltammetry ($\approx$ 10KV·s^{-1}), made possible by lower iR losses at small electrodes, was also demonstrated (2) by Parker.

While 2.8mm Pt electrodes could not detect a concentration polarization current in a benzene solution of perylene containing 0.5$\underline{M}$ HexNClO$_4$, well behaved cyclic voltammograms at sweep rates up to several hundred volts per second could be obtained at Pt ultramicroelectrodes of 30μm diameter. Derivative cyclic voltammetry was performed in more conductive dimethylformamide solution with electrodes of 800μm diameter at sweep rates up to 10KV·s^{-1}, demonstrating the importance of radial mass transport (at the electrode edge) in the reduction of capacitative and ohmic losses in fast transient experiments at electrodes.

References

1. R. Lines and V.D. Parker, Acta Chem. Scand., $\underline{B31}$ (1977) 369.
2. E. Ahlberg and V.D. Parker, Acta Chem. Scand., $\underline{B33}$ (1979) 696.

Stanley Pons

Pons et al. have been concerned with a variety of applications of ultramicroelectrodes to electrochemical experiments. Electrochemistry in the gas phase has been demonstrated with the use of an ultramicroelectrode configuration that allows surface diffusion of ion between the two electrodes (1). The device has been demonstrated as sensitive and species selective gas chromatographic detector. (See Chapter 5 for more information.) Two papers with Fleischmann et al. (2,3) were concerned with the novel use of dispersions as electrodes. When small particles of metals (or semiconductors) are dispersed between plate feeder electrodes in a solution containing no purposely added supporting electrolyte, it becomes possible to drive electrochemical reactions on opposite sides of the particles by applying high voltages at the feeder electrodes. The workers demonstrated the effect by investigating the H^+/H_2 reaction at a platinum dispersion, as well as oxygen reduction in other experiments. There are many new applications to be realized with this type of system. The authors point out that it provides a direct method for the study of catalytic reactions, as well as a convenient way to carry out electrosynthetic reactions at ultramicroelectrodes.

Other work was concerned with the oxidation of species with very high vertical ionization potentials (4). It was shown, for instance, that it is possible to oxidize methane and other aliphatic hydrocarbons in acetonitrile solution that contains no added supporting electrolyte. This was possible since it is the supporting electrolyte that is responsible for the conventional anodic potential limit in that solvent. It was then possible to carry out electrochemical experiments at much greater positive potentials. Indeed, it was shown that it was possible to investigate the electrochemistry of the rare gases and oxygen

and nitrogen in similar solvent systems. These species were
oxidized at potentials less than that predicted from a linear
half wave/ionization potential model due to the difference in
solvation energies between the oxidized and reduced forms of
these small molecules.

Two theoretical papers have been published concerning the
mass transport phenomena to the ring and disk geometries (5,6).
The second of these presents an exact analysis of the problem,
and defines the way that must be considered to obtain the time
dependent response to these geometries. The analyses show that
the very thin ring geometries are to be preferred always over
the disk, since the differences in the results for the cases of
constant flux and constant concentration boundary conditions
are minimized at the thin ring geometry.

Another paper deals with the determination of heterogeneous
rate constants for electron transfer reactions in the steady
state by the use of ultramicroelectrodes. The analysis used was
described in the above theoretical papers. Rate constants up to
a few cm s^{-1} may be determined by the method.

Work has been done on the ferrocene oxidation reaction at
platinum ultramicroelectrodes (7). The results show that under
usual conditions, the ferrocene oxidation is not a simple quasi
reversible reaction as has been previously supposed. Indeed,
the ferrocenium species apparently reacts with the acetonitrile
to form a conducting film on the surface of any sized platinum
electrode. The reaction mechanism is discussed.

Simulation of edge effects to disk electrodes and disk
arrays was the subject of a series of papers by Pons et al.

References

1. J. Ghoroghchian, F. Sarfarazi, T. Dibble, J. Cassidy, J. J. Smith, A. Russell, M. Fleischmann and S. Pons, Analytical Chemistry, 58 (1986) 2278.
2. M. Fleischmann, J. Ghoroghchian, and S. Pons, J. Phys. Chem., 89 (1985) 5530.
3. M. Fleischmann, J. Ghoroghchian, D. R. Rolison, and S. Pons, J. Phys. Chem., 90 (1986) 6392.
4. T. Dibble, S. Bandyopadhyay, J. Ghoroghchian, J. J. Smith, F. Sarfarazi, M. Fleischmann and S. Pons, J. Phys. Chem., 90 (1986) 5275.
5. M. Fleischmann, S. Bandyopadhyay and S. Pons, J. Phys. Chem., 89 (1985) 5537.
6. M. Fleischmann and S. Pons, J. Electroanal. Chem., 222 (1987) 107.
7. J. Daschbach, D. Blackwood, J. W. Pons and S. Pons, J. Electroanal. Chem., in the press.
8. B. Speiser and S. Pons, Can. J. Chem., 60 (1982) 1352.
9. B. Speiser and S. Pons, Can. J. Chem., 60 (1982) 2463.
10. B. Speiser and S. Pons, Can. J. Chem., 61 (1983) 156.
11. J. Cassidy, S. Pons and B. Speiser, Can. J. Chem., 62 (1984) 716.
12. J. Cassidy and S. Pons, Can J. Chem., 63 (1985) 3577.
13. J. Cassidy, J. Ghoroghchian, F. Sarfarazi, J. J. Smith and S. Pons, Electrochim. Acta, 31 (1986) 629.

Attilla Szabo

Attila Szabo and co-workers have been active in the mathematical theory of mass transport to both single and array ultramicroelectrode systems. One area of interest has been the adaptation of the hopscotch finite difference approximation algorithm (1) to the differential equations governing electrochemical problems. The results are generally obtained faster and are more accurate than the conventional explicit finite difference technique, although they are somewhat less accurate than those obtained by the implicit Crank-Nicholson or alternating direction techniques (2,3). Comparisons have not been made with the methods of weighted residuals, which have been shown to have advantages to some systems. One of the simulations applied in the initial work (1,3) was concerned with the finite disk electrode, a problem of major importance to ultramicroelectrode studies. This work was extended (4) to include the chronoamperometric response at the microdisk configuration in a finite insulating disk plane, and also to the problem of chronoamperometry at an array of microdisk electrodes imbedded in an insulating plane (5). This problem has been undertaken by several workers in earlier reports, both in terms of approximate analytical (6,7) and explicit finite difference (8) simulation models. All of these models predict Cottrell behavior at times up to the onset of the building of the "spherical" diffusion layer, and then subsequent to the development of planar diffusion as the adjacent "spherical" fields overlap. The intermediate region has been predicted by several workers with varying degrees of precision, which is a good indication of the difficulty experienced in choosing good models for this geometry (*vide infra*).

Szabo, along with Dennis Tallman, Mark Wightman and co-workers (9) have also approached the theoretical treatment of hemicylindrical and band ultramicroelectrodes. A chronoamperometric result for the band electrode was determined and details are given in Chapter 5. Their results for the ring geometry are also presented there.

References

1. A.R. Gourlay, J. Inst. Math. Appl., $\underline{6}$ (1970) 375.
2. D. Shoup and A. Szabo, J. ELectroanal. Chem., $\underline{140}$ (1982) 237.
3. D. Shoup and A. Szabo, J. Electroanal. Chem., $\underline{160}$ (1984) 1.
4. D. Shoup and A. Szabo, J. Electroanal. Chem., $\underline{160}$ (1984) 27.
5. D. Shoup and A. Szabo, J. Electroanal. Chem., $\underline{160}$ (1984) 19.
6. J. Lindemann and R. Landsberg, J. Electroanal. Chem., $\underline{30}$ (1971) 79.
7. T. Gueshi, K. Tokuda and H. Matsuda, J. Electroanal. Chem., $\underline{89}$ (1978) 247.
8. H. Reller, E. Kirowa-Eisner and E. Gileadi, J. Electroanal. Chem., $\underline{138}$ (1982) 65.
9. A. Szabo, D.K. Cope, D.E. Tallman, P.M. Kovach and R.M. Wightman, J. Electroanal. Chem., $\underline{217}$ (1987) 417.

Dennis Tallman

Dennis Tallman has pointed out that theoretical description at planar unshielded ultramicroelectrodes is complicated by the mixed boundary conditions which exist on the surface containing the electrodes (1). As a result, complete analytical solutions for the diffusion problem are difficult to obtain. Numerical methods on the other hand always present an excellent opportunity for obtaining complete descriptions of ultramicroelectrode behavior. Tallman and co-workers carry out their analysis by transforming the differential equations describing the diffusion problem into integral equations that are then solved numerically. The workers claim superior performance in computational speed over finite difference techniques (1-4). Tallman has applied this and other numerical techniques to a variety of problems, including calculation of the convective-diffusion current at multiple strip electrodes in a rectangular flow channel (1). A spectral method was used to describe this problem. Assuming viscous flow conditions while neglecting edge diffusional processes, the authors assessed the contribution to the enhancement from depletion layer recharge which took place as analyte flowed over the insulating regions between electrodes of the array. Comparing and extending the results found for inviscid flow, he examined the significance of longitudinal edge diffusion at small electrodes and its contribution to the current under convective conditions. The implications of these results were discussed from the point of view of designing electrochemical detectors having higher signal to noise ratio for improved detection in flowing streams.

The integral equation method has been described (2). First the original diffusion equations were Laplace transformed from the time (T) domain, which gives a "stationary" S-domain equation in the Laplace transformed concentration $\hat{C}$. Standard techniques convert this new differential equation into an integral equation with the flux $\partial \hat{C}/\partial N$ as a variable in the integral equation. The integral equation was then solved to obtain the Laplace space current $\hat{I}(s)$, and the current function inverted to time to give the current $I(t)$. The technique requires an explicit knowledge of the Neumann function, and an efficient procedure for its numerical evaluation. As well, it requires a solution method for the integral equation, and a reliable method for the inversion of the Laplace transformed current. This new approach to solving electrochemical diffusion problems was applied to obtain a mathematically accurate (0.1%) description of the time dependence of the diffusion-limited current at a planar band microelectrode; the calculations covered four decades in time. Tallman's results agreed well with the numerical results available from other workers.

Tallman participated in further work involved with obtaining solutions to the problem of chronoamperometry at hemicylinder and band microelectrodes; the reader is referred to the review on Szabo's work.

In later work (4), Tallman investigated the problem of the potential step perturbation at ultramicroelectrodes. He pointed out that even for the simple case of diffusion limited current the theoretical treatment of the problem is complex due to the mixed boundary conditions. Extension of his integral equation method to more general boundary conditions led him to predict simplifications which do not appear to have been previously reported. These require only that the conjugate redox species have equal diffusion coefficients and are applicable to experiments involving a potential step (or sequence of potential steps) at an electrode consisting of one or more active regions of arbitrary geometry in a bounded or unbounded cell of arbitrary geometry. As a result, he was able to show that the reversible electron transfer case can be expressed in terms of a solution for the diffusion limited case, and that a solution for the quasi reversible case can be given in terms of the solution for the totally irreversible case. He demonstrated the method for several geometries, including planar, spherical, and cylindrical.

References

1. D.K. Cope and D. Tallman, J. Electroanal. Chem., 205 (1986) 101.
2. S. Coen, D.K. Cope and D. Tallman, J. Electroanal. Chem., 215 (1986) 29.
3. A. Szabo, D.K. Cope, D. Tallman, P. Kovach and M. Wightman, J. Electroanal. Chem., 217 (1987) 417.
4. D.K. Cope and D. Tallman, J. Electroanal. Chem., in press.

Wolfgang Thormann

Thormann has investigated linear microelectrode arrays prepared by microlithographic techniques and their applications to on-line electric field sensors in electrophoretic capillary columns with special emphasis on measuring the dynamics of the development of potential gradients along the separation axis (1-7), as well as for voltammetry (8). The fabrication principles investigated include: i) the wet chemical etching of vapor deposited thin layers of conductors on glass such as gold, SnO_2 and In_2O_3/SnO_2 (80/20%), and ii) the application of gold or platinum metallo organics by silk screening onto pyrex glass followed by heat decomposition of the organic matrix at 675°C. Arrays with 3 to 256 sensing elements, having electrode widths from 15 to 500μm and spacings from 30 to 920μm have been designed with electrode heights of approximately 0.1μm.

An array of equally spaced ultramicroelectrodes along an electrophoretic capillary column allows migration rates of ionic solutes to be readily determined and electric field profiles to be monitored with high resolution. Other applications of the array detectors comprise the evaluation and criticism of anomalous conductivity zones, including those encountered in gel-solution interfaces. In the area of moving boundary electrophoresis, boundary location, boundary structure, and length of zones have all been monitored as a function of time. In isotachophoresis, Thormann has investigated monitoring the separation process explicitly, i.e., in terms of the time and location of attaining a migrating steady state. Also, the continuous broadening of a migrating solute in zone electrophoresis has been visualized. An array ultramicroelectrode detector can be controlled with a microprocessor system. In the isotacophoretic approach, the potential gradient was measured at 255 equidistant points along the separation axis. Each sensor was 60μm in width and had a center separation distance of 340μm. SnO_2 (as opposed to Au or In_2O_3/SnO_2 (80%/20%) electrodes were found to be the most reliable and sensitive. For instance, the isotachopherograms obtained in the separation of maleic and phosphoric acid demonstrated the impressive ability of the detector to respond to small changes of the electric field across an isotachophoretic zone structure. Application to the automation of analytical isotachophoresis, and time domain analysis of moving boundaries and migrating sample zones have been published (1-4).

Isoelectric focusing was cited as a prime area of investigation with ultramicroelectrode arrays. Specifically, Thormann described techniques whereby information regarding the attainment of the stationary steady state zone pattern and the electric field strengths across focused protein bands could be determined. The latter, in effect, will allow the determination of the diffusion coefficients and other physical parameters of proteins. Problems commonly encountered with the technique, including drifts along the focusing column, can be readily probed in the time domain with ultramicroelectrodes. The potential gradient in these experiments was measured along a 10cm focusing column using 100 sensing gold electrodes fabricated from metallo organics (5-7).

For voltammetry, the gold and platinum arrays have been converted into linear arrays of microband electrodes with individual elements which are geometrically equivalent to circular disk ultramicroelectrodes of radii 0.6 to 5μm. Voltammetric measurements can be made at individual sensing elements, part of the array, or the total linear ensemble. The response of the array is close, but not exactly equal, to the sum of the individual sensing elements. A steady state current is yielded with currents in the μA range (1m$\underline{M}$ ferrocene in acetonitrile) for the larger arrays. The constructed ensembles exhibit microelectrode characteristics that permit measurements to be made in highly resistive media. Also, the application of transient techniques such as AC and pulse voltammetry with ms pulse widths was investigated in systems with supporting electrolyte and with high resistance (8).

Other areas of research have been: i) the fabrication, theory and voltammetric characterization of single and arrays of microline (band), microring and microdisk electrodes (9,10); ii) the assessment of fabrication quality of these electrodes using transient electrochemical techniques (11); iii) the elucidation of the very fast electron transfer rate for oxidation of ferrocene at ultramicroelectrodes (10); iv) examination of staircase, normal pulse and differential pulse voltammetry at ultramicroelectrodes using a computerized multi-time domain measurement method and different electrode formats (12); and v) the use of microelectrodes for electroanalysis in static and flow through configurations (13).

References

1. E. Schumacher, W. Thormann and D. Arn, in Analytical Isotachophoresis, F.M. Everaerts, ed., Elsevier, Amsterdam, 1981, pp. 33.
2. W. Thormann, D. Arn and E. Schumacher, Separation Science and Technology, 19 (1984-85) 995.
3. W. Thormann, D. Arn and E. Schumacher, Electrophoresis, 5 (1984) 323.
4. W. Thormann, J. Chromat., 334 (1985) 83.
5. W. Thormann, G. Twitty, A. Tsai and M. Bier, in Electrophoresis '84, V. Neuhoff, ed., Verlag Chemie, Weinheim, 1984, pp. 114.
6. W. Thormann, N.B. Egen, R.A. Mosher and M. Bier, J. Biochem. Biophysics. Meth., 11 (1985) 287.
7. W. Thormann, R.A. Mosher and M. Bier, J. Chromatogr., 351 (1986) 17.
8. W. Thormann, P. van den Bosch and A.M. Bond, Anal. Chem., 57 (1985) 2764.
9. A.M. Bond, T.L.E. Henderson and W. Thormann, J. Phys. Chem., 90 (1986) 2911.
10. A.M. Bond, T.L.E. Henderson, T.F. Mann, D.R. Mann, W. Thormann and C.G. Zoski, Anal. Chem., submitted for publication.
11. W. Thormann and A.M. Bond, J. Electroanal. Chem., 218 (1987) 187.
12. W. Thormann, J.W. Bixler, T. Mann and A.M. Bond, Pittsburgh Conference 1987, Atlantic City, abstract #181.
13. J.W. Bixler, A.M. Bond, P.A. Lay, W. Thormann, P. van den Bosch, M. Fleischmann and S. Pons, Anal. Chim. Acta, 187 (1986) 67.

Mark Wightman

Mark Wightman has been involved for several years in the development of the field of ultramicroelectrodes. One should consult several other reviews in this volume for other work in which Wightman has been involved. These include, for example, Ewing, Howell, Amatore, Tallman, and Szabo. We will present here a cross section of Wightman's work.

A well known work was Wightman's A-page article in *Analytical Chemistry* in 1981 (1) where he described the analytical usefulness of the devices.

In 1980, Wightman et al. (2) described voltammetric electrochemical behavior at carbon fiber electrodes. The electrodes were used to observe the electrooxidation of dopamine, dihydroxyphenylacetic acid, and ascorbic acid. The results were compared to those predicted by the equations for a hemispherical electrode of identical radius. Equations were derived by evaluating the pertinent flux equations in spherical coordinates. Charge-transfer rates deduced from this analysis were found to be the same at carbon fiber electrodes as at carbon paste for dopamine oxidation. The quasi-reversible behavior of dihydroxyphenylacetic acid at carbon paste electrodes was observed to be enhanced at the fiber electrode. Because of the small size of the carbon fibers, enhanced current from the catalytic oxidation of ascorbic acid by oxidized dopamine was minimized.

Carbon fibers were used in the fabrication of voltammetric electrodes with an active area of approximately 5×10^{-7} cm^{-2} (3). The response of these electrodes was evaluated in aqueous solutions of $K_3Fe(CN)_6$ for cyclic voltammetric, differential pulse voltammetric, and chronoamperometric experiments. The authors claimed that since the diameter of these electrodes is smaller than the characteristic distance for molecular diffusion

on the time scale of the experiments, the current was
essentially time independent. The electrodes were shown to be
useful for analytical determination of concentration for over 3
orders of magnitude.

In 1981, Wightman and co-workers (4) demonstrated that
microvoltammetric electrodes could be used to provide
reproducible results in a complex chemical medium, the mammalian
brain. Because of the small size of the microelectrodes,
current on the second step of a double potential step experiment
was considered to be nonfaradaic and was used for residual
current correction. The use of normal pulse voltammetry
minimized surface adsorption by electrogenerated products. The
response of the electrodes *in vivo* was described to be superior
to that of other electrodes. Their integrity was not altered by
the repetition rate of the potential pulses, and dopamine, a
compound of prime neurochemical interest, could be partially
resolved from ascorbic acid and dihydroxyphenylacetic acid by *in
vivo* voltammetry.

Wightman and co-workers have, since that time, completed a
number of ultramicroelectrode biochemical studies involving the
mammalian brain (5-11).

Ultramicrovoltammetric electrodes were examined (12,13) in
ultrafast voltammetric experiments. The workers examined the
oxidation of anthracene and 9,10-diphenylanthracene in aprotic
solvents. Decreased iR drop was deserved in the oxidation of
ferrocene in acetonitrile, methylene chloride, tetrahydrofuran,
dimethoxyethane, chlorobenzene, and benzene.

A series of papers by Wightman and co-workers (14-21) was
concerned with the theory and applications of band and cylinder
ultramicroelectrodes in analytical chemistry. This area is
covered in detail in other chapters of this volume.

These workers have also studied the response at plated

mercury ultramicroelectrodes (22). A method for the preparation of these devices with radii of 2.3-7.3μm was described. Mercury was electrodeposited from solutions of Hg(I) onto a microvoltammetric platinum disk electrode at a constant potential. The radius of the deposited mercury electrode was predicted to be a function of the square root of the deposition time and was experimentally confirmed at a hemispherical mercury electrode in the reduction of $Ru(NH_3)_6^{3+}$. The electrodes were also used to demonstrate cyclic voltammetry and stripping voltammetry.

Wightman et al. (23) have discussed the proper construction of gold disk electrodes for minimization of cracking and leakage at the tip, which contributes to the capacitative noise at low voltammetric scan rates. The workers hydrophobically treated cracks by a silanization procedure. Marked reduction in capacitative currents were observed.

The workers have also published work (24) on background current subtraction for rapid scan voltammetry. This work is discussed in the review of Howell's work.

References

1. R.M. Wightman, Anal. Chem., 53 (1981) 1125A.
2. M.A. Dayton, A.G. Ewing and R.M Wightman, Anal. Chem., 52 (1980) 2392.
3. M.A. Dayton, J.C. Brown, K.J. Stutts and R.M Wightman, Anal. Chem., 52 (1980) 946.
4. A.G. Ewing, M.A. Dayton and R.M. Wightman, Anal. Chem., 53 (1981) 1842.
5. M.A. Dayton, A.G. Ewing and R.M. Wightman, J. Electroanal. Chem., 146 (1983) 189.
6. A.G. Ewing, J.C. Bigelow and R.M. Wightman, Science, 221 (1983) 169.
7. P.M. Kovach, A.G. Ewing, R.L. Wilson and R.M. Wightman, J. Neuroscience Meth., 10 (1984) 215.
8. J. Millar, J.A. Stamford, Z.L. Kruk and R.M. Wightman, Eur. J. Pharm., 109 (1985) 341.
9. W.G. Kuhr and R.M. Wightman, Brain Res., 381 (1986) 168.
10. C. Amatore, R.S. Kelly, E.W. Kristensen, W.G. Kuhr and R.M. Wightman, J. Electroanal. Chem., 213 (1986) 31.
11. R.S. Kelly and R.M. Wightman, Anal. Chim. Acta, 187 (1986) 79.
12. J.O. Howell and R.M. Wightman, J. Phys. Chem., 88 (1984) 3915.
13. J.O. Howell and R.M. Wightman, Anal. Chem., 56 (1984) 524.
14. P.M. Kovach, W.L. Caudill, D.G. Peters and R.M. Wightman, J. Electroanal. Chem., 185 (1985) 285.
15. K.W. Wehmeyer, M.R. Deakin and R.M. Wightman, Anal. Chem., 57 (1985) 1913.
16. C.A. Amatore, M.R. Deakin and R.M. Wightman, J. Electroanal. Chem., 206 (1986) 23.
17. P.M. Kovach, M.R. Deakin and R.M. Wightman, J. Phys. Chem., 90 (1986) 4612.
18. M.R. Deakin, R.M. Wightman and C.A. Amatore, J. Electroanal. Chem., 215 (1986) 49.
19. A. Szabo, D.K. Cope, D.E. Tallman, P.M. Kovach and R.M. Wightman, J. Electroanal. Chem., 217 (1987) 417.
20. C.A. Amatore, B. Fosset, M.R. Deakin and R.M. Wightman, J. Electroanal. Chem., submitted.
21. C.A. Amatore, M.R. Deakin and R.M. Wightman, J. Electroanal. Chem., submitted.
22. K.R. Wehmeyer and R.M. Wightman, Anal. Chem., 57 (1985) 1989.
23. K.R. Wehmeyer and R.M. Wightman, J. Electroanal. Chem., 196 (1985) 417.
24. J.O. Howell, W.G. Kuhr, R.E. Ensman and R.M. Wightman, J. Electroanal. Chem., 209 (1986) 77.

5
Contributions

This chapter contains contributions that were made at The Utah Workshop and Conference on Ultramicroelectrodes held at Midway, Utah, in late January, 1986. The meeting was organized around four aspects of the field of ultramicroelectrodes: (1) Theoretical, (2) Systems, (3) Analytical, and (4) Construction.

After general discussion, presentations of new research results were offered in each of the areas by all of the participants.

The contributions herein reflect the intensity of research in the field and the innovative efforts of the contributors.

H.D. Abruña contributed the following on analytical chemistry with ultramicroelectrodes.

We have been involved in the development of electrodes modified with functionalized polymer films for the determination of metal ions in solution (1-3). The concept, schematically depicted in Figure 1, is based on the modification of the surface of an electrode with a polymer film that contains both a ligand group, chosen so as to have high affinity and selectivity for a given metal ion, as well as an additional redox couple that serves as an internal standard which allows one to have an *a priori* knowledge of the surface coverage and aids in the deposition of the polymer.

The ligand can be either part of the polymer backbone or can be incorporated, via ion exchange, into a polycationic film. The former gives films of superior stability whereas the latter is more versatile due to the large number of anionically (usually through a sulfonate side chain) charged ligands.

The modified electrode is exposed to the analysis solution where coordination of the ion from solution to the surface immobilized ligand takes place. Once coordinated, the electrochemical response of the immobilized metal/ligand complex is used as the analytical signal which is related to the concentration of the metal ion in solution.

Conceptually, this is analogous to stripping voltammetry except that the preconcentration step is achieved via coordination and not by electrolysis. This allows for the use of selectivity trends exhibited in solution as a means of selecting the best reagent for a particular application and can in principle, through the appropriate choice of reagents, give rise to unprecedented levels of selectivity and sensitivity.

We have demonstrated the feasibility of this approach to the

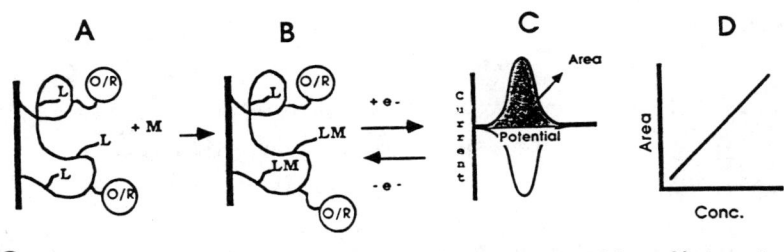

Figure 1. Schematic depiction of the use of polymer modified electrodes for the determination of metal ions in solution.

determination of metal ions in solution, even with oxidation state selectivity. This approach has also been employed by other investigators (4-8). Most recently, we extended the applicability of this technique to the determination of organic functionalities (9).

In addition to the general analytical utility of these investigations, we are also interested in the development of electrochemical microprobes for performing analytical determinations in very small samples including single cell specimens. Of particular interest is the determination of Ca^{2+} and Mg^{2+} due to their great relevance in numerous biological functions. These and other ions are present in intracellular fluids at typically greater than $\mu\underline{M}$ levels; these fall well within the limits of detection that we have achieved at conventionally sized electrodes. In addition, the amperometric mode of detection is attractive since it has great sensitivity and is relatively immune to potential drifts, a problem that plagues potentiometric measurements.

When employing microelectrodes, it is important to determine whether there will be a degradation in the signal to noise (S/N) ratio and whether the expected small currents will be measurable. With this in mind, we have extended our studies on metal ion analysis to platinum microelectrodes (25μm diameter). The system that we have employed is a copolymer of $[Os(v\text{-}bpy)_3]^{+2}$ (v-bpy is 4-vinyl,4'-methyl-2,2'-bipyridine) with v-bpy in a 1:6 mole ratio. Previous studies with conventionally sized electrodes had indicated that this polymer adsorbs onto platinum electrodes and that the immobilized ligand readily coordinates Fe(II) from solutions as dilute as $10^{-7}\underline{M}$. The osmium complex within the polymer has a reversible redox response at about +0.72V whereas that for the iron complex is at about +1.0V so that their responses are well separated.

Figure 2 presents a differential pulse voltammogram for a modified platinum electrode prior to and after exposure to a 8×10^{-6} M solution of Fe(II). In curve A one can observe a well developed response for the $[Os(vbpy)_3]^{+2}$ couple within the polymer and essentially background current out to +1.3V. After exposure to the Fe(II) solution, a well developed and easily quantifiable signal (curve B) for the surface immobilized $[Fe(vbpy)_3]^{+2}$ can be observed at about +1.0V. In addition to differential pulse voltammetry (DPV), we also investigated the applicability of square wave voltammetry to these studies due to its high sensitivity and short experimental times as compared to DPV. Figure 3 shows forward (A), reverse (B) and difference (C) currents for a modified platinum microelectrode after exposure to a 1×10^{-6} M solution of Fe(II). Again, very well developed signals could be obtained. It should be noted that a monolayer of electroactive material at these electrodes represents about 5×10^{-16} moles!

Having established the feasibility of employing modified microelectrodes for analytical determinations we proceeded to identify materials that could be employed in the determination of Ca^{2+} with high sensitivity and selectivity. The reagent chosen was antipyrylazo III(AP-III), shown in Figure 4A. This was incorporated by ion exchange onto a platinum electrode (of conventional size in these studies) modified with a film of polybenzylviologen (PBV) (Figure 4B).

The electrode modified with the viologen polymer and AP-III exhibits two reversible reductions at negative potentials (Figure 5A) due to the viologen centers. The AP-III exhibits an irreversible oxidation wave at about +0.55V (Figure 5B) the height of which is attenuated after coordination of Ca^{+2}. The peak current value for this process is employed as the analytical signal and a calibration curve such as that shown in

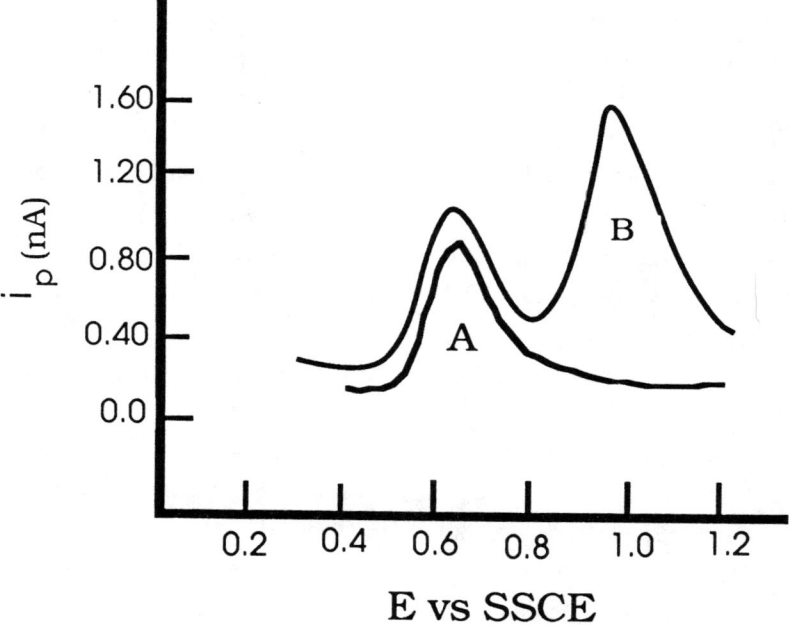

Figure 2. Differential pulse voltammogram for a 25 micron diameter platinum electrode modified with a $[Os(v\text{-}bpy)_3]^{+2}/v\text{-}bpy$ (1:6) copolymer prior to (A) and after (B) exposure to a $6\times10^{-8}\underline{M}$ solution of Fe(II).

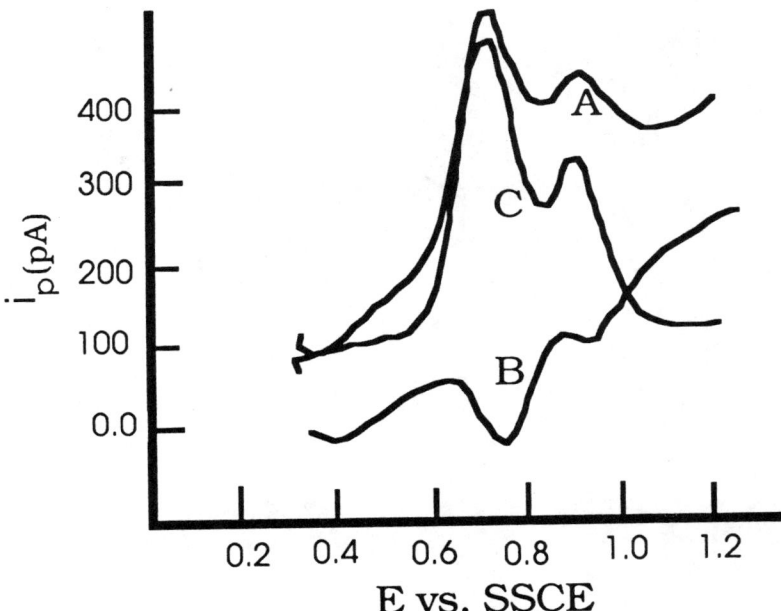

Figure 3. Forward (A); reverse (B); and difference (C) currents of a square wave voltammogram for a platinum electrode modified as in Figure 2 and after exposure to a $1 \times 10^{-6}\underline{M}$ solution of Fe(II).

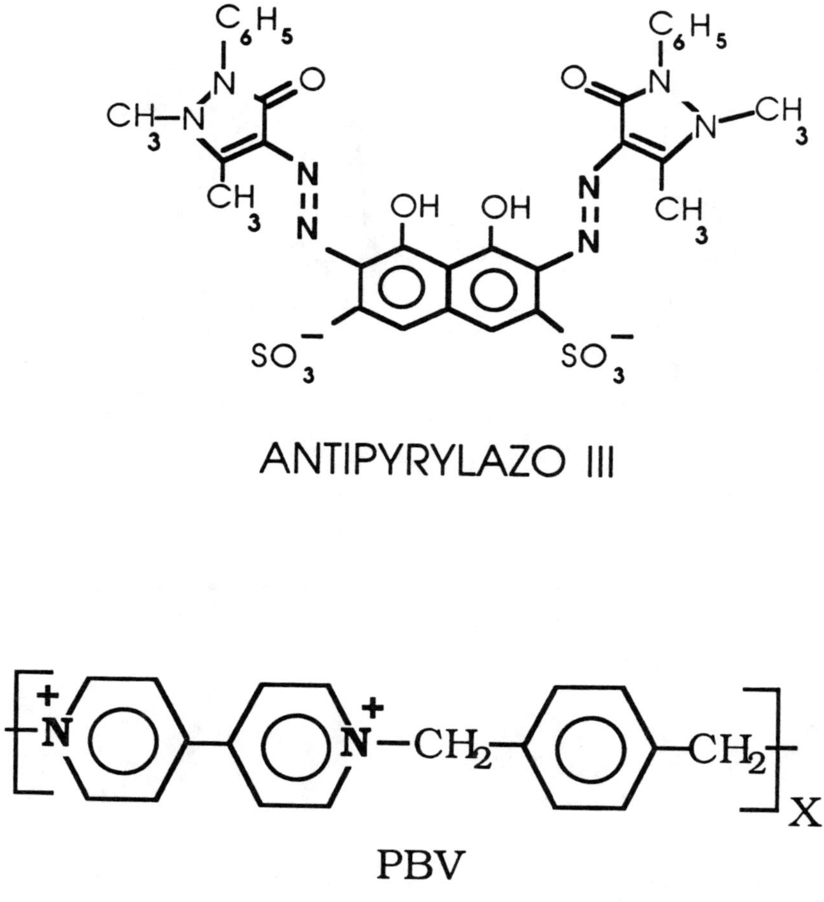

Figure 4. A. Structure of antipyrylazo III (AP-III)
B. Structure of viologen polymer (PBV)

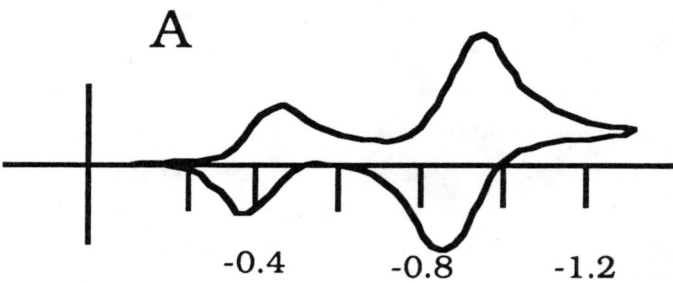

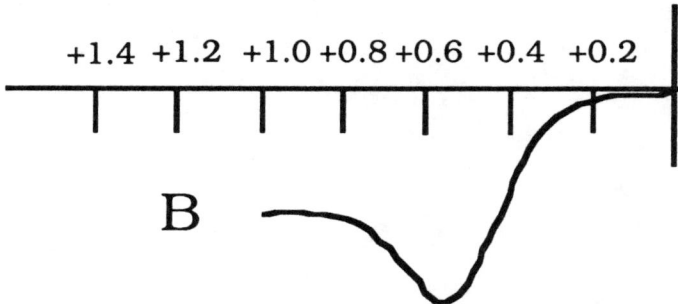

Figure 5. A. Cyclic voltammogram (cathodic sweep) for a platinum electrode modified with PBV and AP-III.
B. Differential pulse voltammogram for the electrode in Figure 5A depicting the irreversible oxidation of AP-III bound to the electrode.

Figure 6 can be obtained. It is important to note that not only is the technique sensitive to calcium, but also the addition of even a thousand-fold excess of magnesium (traditionally the most severe interferent in calcium determinations) has a relatively minor effect. This is shown in the calibration curve in Figure 6 where the signal due to calcium at $1 \times 10^{-6} \underline{M}$ concentration was only slightly perturbed by the presence of magnesium at millimolar concentration.

When combined with the previously described work on microelectrodes, it is clear that this represents a viable approach to the determination of intracellular levels of numerous metal ions. We are currently developing strategies for the preparation of electrodes of submicrometer total diameter so that they can be implanted into single cells without damaging it. We are confident that this approach will open new venues for the investigation of intracellular levels of metal ions.

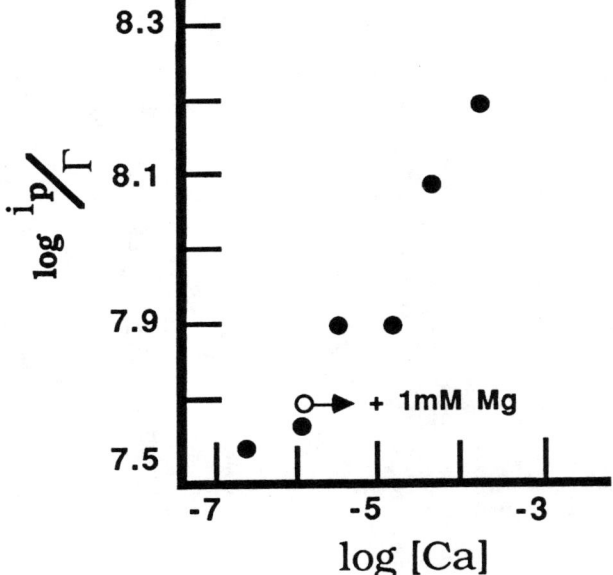

Figure 6. Calibration curve for the determination of Ca^{+2} with an electrode modified with PBV and AP-III. Open circle: response for a $10^{-6}\underline{M}$ solution of calcium after addition of Mg at the $1 \times 10^{-3}\underline{M}$ level.

Acknowledgements

This work was generously supported by the National Science Foundation, Dow Chemical Co., Eastman Kodak Co. and Honeywell Inc. The co-author of this work, H. C. Hurrell acknowledges support by a fellowship from the Aerospace Corporation. H.D.A. acknowledges support via the Presidential Young Investigator Award Program of the National Science Foundation.

References

1. A.R. Guadalupe and H.D. Abruña, Anal.Chem., 57 (1985) 142.
2. L.M. Wier, A.R. Guadalupe and H.D. Abruña, Anal. Chem., 57 (1985) 2009.
3. A.R. Guadalupe, L.M. Wier and H.D. Abruña, American Laboratory, 18 (1986) 102.
4. J.A. Cox and M. Majda, Anal. Chem., 52 (1980) 861.
5. M.J. Gerhon and A. Brajter-Toth, Anal. Chem., 58 (1986) 1488.
6. D.J. Harrison, private communication.
7. R.P. Baldwin, J.K. Christensen and L. Kryger, Anal. Chem., 58 (1986) 1790.
8. G.T. Cheek and R.F. Nelson, Anal. Lett., 11 (1978) 393.
9. A.R. Guadalupe and H.D. Abruña, Anal. Lett., 19 (1986) 1613.

Christian Amatore presented the following lecture on the theoretical considerations of mass transport to, and kinetic processes at ultramicroelectrodes.

The first part of this presentation is devoted to a brief summary of the properties of ultramicroelectrodes and to the experimental applications which may be inferred from these properties. This in turn allows a description of the theoretical treatments and design of these experiments.

The second part of the presentation presents some theoretical approaches recently developed by this author in collaboration with Professor Wightman.

Some Properties of Ultramicroelectrodes and Their Implications

All the properties of ultramicroelectrodes are obviously related to their small size. As will be seen in the following treatment, this small size has a number of consequences which can be considered sequentially for the sake of simplicity. Note however that these consequences are usually related.

Characteristic dimension

This notion will not be developed here since it has been extensively treated in Professor Oldham's lecture (*vide infra*). Let us simply denote here r_0 as this characteristic dimension/size. The main feature of ultramicroelectrodes is that r_0 is much smaller than δ_{conv}, the thickness of the convection layer. Thus this allows us to make the diffusion layer, δ, larger than r_0, without any interference from convection.

Conversely when $r_0 < \delta < \delta_{conv}$ or $r_0 \approx \delta < \delta_{conv}$, edge effects will increase and can even predominate. Thus nonplanar diffusion is observed and the symmetry of the diffusion field is now governed by the very shape of the electrode. Problems related to nonuniform current densities may arise in this situation (*vide infra*).

Such a situation clearly warrants the derivation of relationships characterizing the behavior of each type of electrode, i.e., each electrode shape. Unfortunately the most convenient geometries to produce experimentally (i.e., disk, band, ring ...) are not those most conveniently described mathematically. Thus, methods are needed for the derivation of theoretical laws at realistic geometries; some methods already exist but usually lead to sophisticated analytical treatments or large amounts of time and memory occupation in computers. This problem is even more severe when kinetic complications are taken into account.

Let us consider for example the simple EC scheme when the rate constant k

$$A + e \rightleftarrows B$$

$$B \xrightarrow{k} C$$

is large enough for the associated kinetic layer to be smaller than r_o: $\delta_{kin} \ll r_o < \delta$. It is seen that while A and C concentrations will diffuse under nonplanar conditions, that of B obeys planar diffusion laws. This is clearly a more severe problem to solve than the case where $k = 0$.

High current densities

The enhancement of mass transport at ultramicroelectrodes occurs because either the small size of the diffusion layer (when $\delta \ll r_o$) or lateral diffusion (when $\delta > r_o$ or $\delta \approx r_o$) allows more material to be brought to the electrode. Then, although the number of volume elements flowing to these electrodes is very small (nA or pA range) the resulting current densities may reach several A cm^{-2}. This is larger than observed in many industrial electrolytic processes. This is clearly an important consideration for charge transport as well as for energy dissipation at the very tip of the electrode.

Nonuniform current densities

Edge effects play an important role in diffusion to microelectrodes when $\delta > r_o$ or $\delta \approx r_o$. As a result, for most geometries (e.g., disk, band, ring ...) nonuniform current densities are observed. For a constant surface concentration boundary condition where the electrode and insulator are in the same place, infinite flux will occur at the edge. However, the interference of (i) control by charge transfer and/or (ii) ohmic drop, will decrease these effects. Thus it is seen that the current distribution at the surface of the electrode will be affected not only by mass transport, but will be smoothed (equalized) by other contributions such as (i) or (ii).

Ohmic drop

Electrochemistry at ultramicroelectrodes results generally in little ohmic contribution in the experiment. Indeed ohmic drop is predicted to become vanishingly small when the characteristic dimension of the electrode is made very small (at least if uniform current distribution is considered and when δ does not tend to zero).

This allows electrochemical experiments to be performed in highly resistive media such as in solutions of poor conductivity and low dielectric constant. The use of ultramicroelectrodes also allows electrochemists to study reaction mechanisms and chemical reactivities in conditions close to those found in conventional organic and inorganic synthetic chemistry. However such a situation is clearly idealistic since in such media the increase of resistivity will tend to compensate the gain due to the geometrical factors of the resistance. Effects arising from migration must be taken into account in the derivation of theoretical treatments of diffusion at ultramicroelectrodes. The near disappearance of ohmic drop contributions allows the use of transient methods in a time domain at least 100 times faster than at normal sized electrodes. Indeed cyclic voltammetric scan rates as large as $10,000 V \cdot s^{-1}$ are permitted without serious complication due to ohmic drop or capacitive currents (J. Howell's review) in conductive media. Scan rates in the range of $1,000,000 V \cdot s^{-1}$ may be achieved in the near future if ultramicroelectrodes are coupled to techniques for elimination of ohmic drop and capacitive effects. This will enable the routine determination of any second order rate constants and of first order rate constants up to $5 \times 10^2 s^{-1}$ (for $10^4 V \cdot s^{-1}$) or $5 \times 10^7 s^{-1}$ (for $10^6 V \cdot s^{-1}$). However at such high scan rates, the diffuse layer will not be negligible when compared to the diffusion layer. Again coupling of migration with diffusion must be considered.

Arrays

Diffusion to arrays of ultramicroelectrodes can be viewed in two limiting cases defined by the eventual overlap of the diffusion layer originating from each individual electrode in the array. When there is no overlap, the array behaves as the sum of its individual components. In this case the device is not an array but an artificial device to increase the overall current flowing through the system. It must be noted that when considering ohmic drop, the array may behave differently from what is simply obtained by summing the currents observed for each active center alone. This effect which may arise in poorly conducting media is due to the fact that although the diffusion is limited around each active center, the current flows through all the cell. Thus the flux tubes will overlap leading to an increased overall resistance when all the centers are working together.

From the schematic representation (Figure 1) it is easily understood that the ohmic drop at the array will be more important than that observed at a single element of the array. This is due to the overlap of the current tubes originating from each component of the array.

Conversely when the diffusion layer originating from each center overlaps, the array has to be considered as a whole electrode and cannot be reduced to one of its individual components. However for arrays containing an infinite number of individual active centers, the edge effects on the border of the array may play a negligible part and owing to symmetry considerations the array may be reduced to one of its individual components for theoretical description.

For arrays with a small number of components the electrochemical behavior of the arrays has to be derived by taking into consideration the whole assembly.

Small geometric size

Since an ultramicroelectrode can be made in the range of a few μm (including the insulating material) they are suited for the use as sensors in small volume detectors. This has already been applied successfully for capillary column detectors (see e.g., Bixler contribution), *in vivo* sensors (Wightman, Ewing). Clearly such applications warrant the development of the proper theory for each type of sensor. Indeed these sensors will lead to the need to consider diffusion coupling with flowing streams

PRIMARY DISTRIBUTION

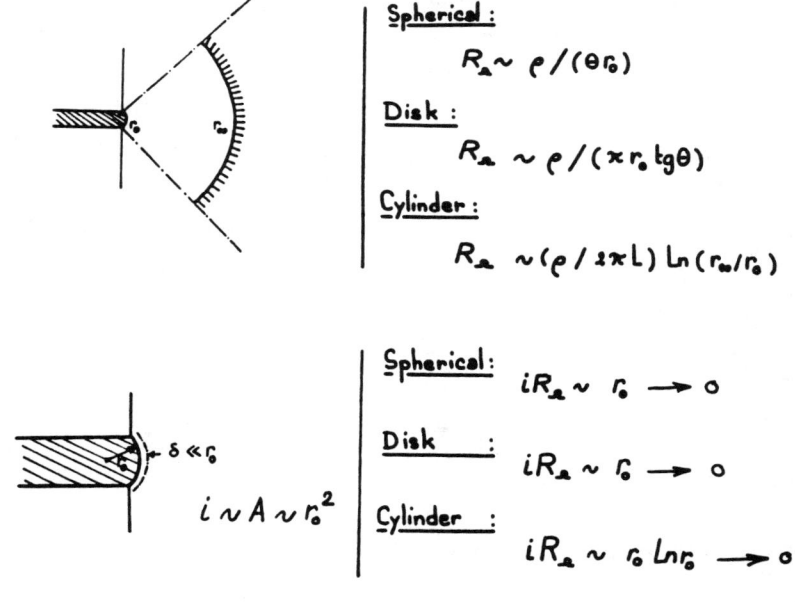

Figure 1. Ohmic drop on the basis of primary current distribution at an ultramicroelectrode under planar diffusion, for three geometries of the electrode: spherical, disk and cylinder.

(HPLC detectors, *in vivo* detection in body fluids, ...) and ion diffusion in anisotropic media (selective coatings on the electrode — Wightman, Abruña — or *in vivo* implantation of electrodes in living tissues — Wightman —).

The Behavior of Ultramicroelectrodes Chemically Implanted in Living Tissue

Wightman gave a review of this problem in his presentation. I will thus simply address here the following question: "When an ultramicroelectrode is chemically implanted, e.g., in brain tissue, does it probe the actual brain tissue, or does it probe only the region where the disruption of the tissue (created by the electrode implantation) predominates?" In other words does this technique of *in vivo* detection probe the properties of the living tissue or some artifact? In order to answer the question we decided to investigate the electrochemical behavior of an electrode actually implanted in brain tissue. In the model (Figure 2), the disruption created to the brain tissue is considered to form a pool (radius R_0) around the tip of the electrode (radius r_0). We note that this model was originally suggested by Adams. For the theoretical derivation, the model was assumed to be of spherical symmetry. Outside of the pool, the brain tissue is assumed to behave as an inert material with reduced partial volume, K, (volume of the extra cellular fluid vs the total volume of the brain tissue) and reduced diffusion coefficient.

To conform with the experimental requirements, a long time approximation was developed (i.e., for times in which steady state currents predominate for the electrode considered). It was shown that the overall phenomenon observed is identical to that which will be observed at an electrode of radius equal to

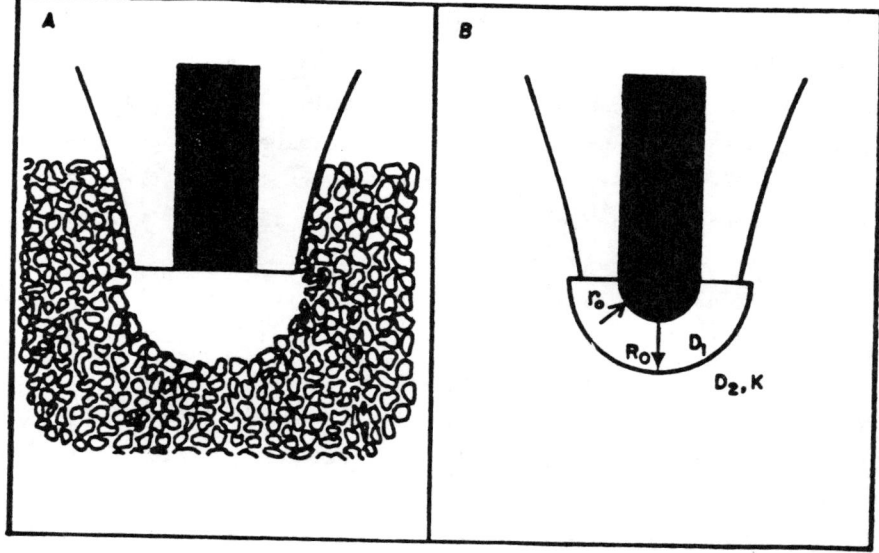

$$k^{o,app.} = k^o \, (r_o/R_o)^2 \, K^{-1}$$

Figure 2. Model for chemically implanted electrode in brain tissue.

the outer radius of the pool, but with an apparent rate constant
of electron transfer at this pseudo electrode:

$$k^{o,app} = k^o \left(\frac{r_o}{R_o}\right)^2 \frac{1}{\kappa}$$

where k^o is the standard rate constant observed in an homogeneous medium at the ultramicroelectrodes of radius r_o, and κ is the relative fraction of the intracellular fluid respective to that of the brain tissue. Good agreement of this model with experimental data was obtained, allowing the determination of $R_o \simeq 10\mu m$ (for $r_o = 5\mu m$) and $\kappa \simeq 15\%$ (Figure 3). The latter value is consistent with $\kappa \simeq 21\%$ determined independently by long range diffusion of Na^+ in the brain.

Electrochemistry at Cylindrical and Band Electrodes
Cyclic Voltammetry

When the length of the electrode is much larger than the diffusion layer, diffusion at cylindrical and band electrodes can be restricted to the treatment of diffusion in a plane perpendicular to the axis of the electrode (Figure 4). Due to the possible transposition of the method discussed to other electrode geometries, we wanted to use a finite difference approach to the problem rather than an analytical one. Indeed the flexibility of finite difference methods allows the consideration of boundary conditions (at the electrode surface) which is hardly possible by analytical methods. However when a finite difference approach is considered for a nonplanar electrode, the usual method requires increasing the number of space elements as the distance from the electrode is increased. This is basically related to the intrinsic concentration

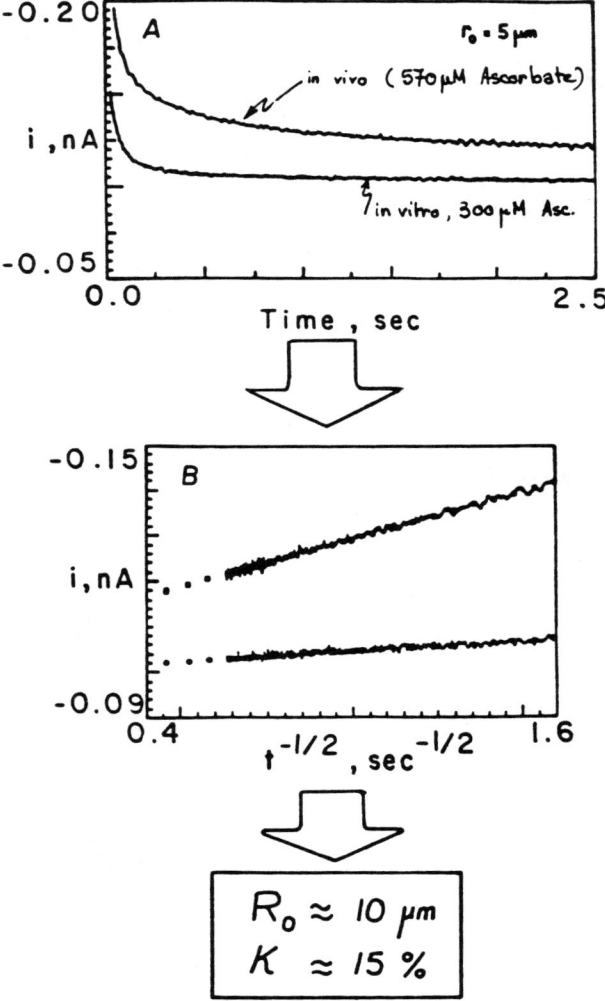

Figure 3. Experimental results for a chemically implanted electrode in brain tissue. Detection of ascorbate.

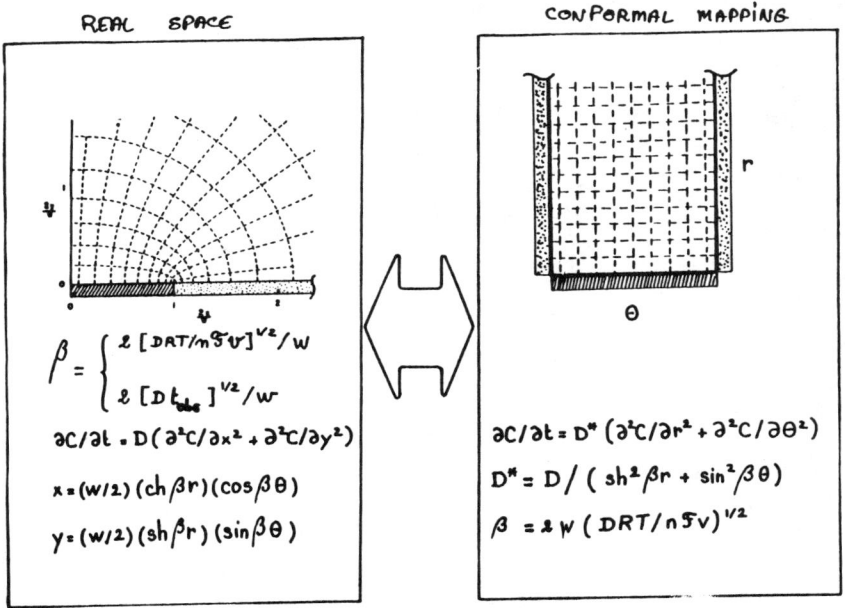

Figure 4. Conformal mapping of the diffusion field at a band electrode.

profiles obtained under such conditions. Our approach is to perform the actual finite difference in a modified space domain which conforms with the properties of the diffusional fields. Mathematical operations allowing such changes are well known (conformal mapping). For example for the band problem, the usual differential equation,

$$\frac{\partial c}{\partial t} = D\left(\frac{\partial^2 c}{\partial x^2} + \frac{\partial^2 c}{\partial y^2}\right)$$

obtained when the "experimental" space domain is considered, is converted to:

$$\frac{\partial c}{\partial t} = D^*\left(\frac{\partial^2 c}{\partial r^2} + \frac{\partial^2 c}{\partial \theta^2}\right)$$

with $D^* = D/(sh^2\beta r - \sin^2\beta\theta)$, $\beta = (2/\omega)$ $(DRT/nF\nu)^{1/2}$ for CV. $\beta = 2\omega(Dt_{obs})^{1/2}$ for chronoamperometry at time t_{obs} ν, ω is the width of the band, and r and θ define the "conformal mapping" of the space as given by the implicit relations:

$x = (\omega/2)$ ch βr cos $\beta\theta$
$y = (\omega/2)$ sh βr sin $\beta\theta$

It is seen in Figure 4 that in the (r,θ) space the diffusion field is closely represented by orthogonal lines to the electrode which simplifies the finite difference solution of the diffusion equation (i.e., less "boxes" are needed for the definition of the grid to get a comparable accuracy). This conformal mapping along with the Hopscotch method allowed the construction of a fast algorithm for determining the concentration profiles.

A similar approach has been taken for the diffusion at a cylindrical electrode. Based on symmetry considerations, we find the problem is only dependent on the distance, r, from the center of the electrode, and the corresponding diffusion equation is:

$$\frac{\partial C}{\partial t} = D \left(\frac{\partial^2 C}{\partial r^2} + \frac{1}{r} \frac{\partial C}{\partial \theta} \right)$$

The introduction of $y = 1/2\ \beta)\ \ln\ (r/r_o)$ where $\beta = (2/r_o)(2RT/nF\nu)^{1/2}$ allowed the above expression to be re-expressed as

$$\frac{\partial C}{\partial t} = D^* \left(\frac{\partial^2 C}{\partial y^2} \right)$$

with $D^* = D\ \exp(-\beta y)$. In the conformal space the $C(y)$ concentration profile is tantamount to those obtained assuming the Nernst diffusion layer approximation. However for large values of β, the apparent thickness, δ_y, of the diffusion layer is dependent on the time elapsed. A proper derivation under these conditions shows that the variation of the latter has a logarithmic dependence with time, i.e., that a quasi-steady state is observed rather than a true steady state such as that observed at a disk or a spherical electrode.

Comparison between the band and cylinder geometry results allowed us to define a correspondence between the two series of results. Indeed, within a reasonable accuracy, the results obtained at a band electrode can be fitted with a cylinder model assuming a relationship between the half width of the band, $\omega/2$, and the radius, r_o, of the equivalent cylinder. However, the relation between $\omega/2$ and r_0 is dependent on β. Current work is in progress to establish this relationship.

Conclusion

The above presentation has tried to define some areas in which theoretical treatments relative to ultramicroelectrodes are clearly warranted. This is not an exhaustive list, but instead presents a series of important problems to be solved if quantitative analysis (kinetics) of data obtained at ultramicroelectrodes is to be realized. This is particularly important due to the fascinating areas being opened to electrochemistry by ultramicroelectrodes (electrochemistry in media of poor conductivity, as well as in gases as discussed by Pons). Clearly such theoretical derivations are also warranted for the optimization of sensors based on ultramicroelectrodes. However, most of the theoretical derivations, as well as those presented in this lecture, involve drastic conditions of symmetry on the electrode shape (i.e., *true* sphere, disk, cylinder, band, ...). In practice, as soon as the use of ultramicroelectrodes is broadened, perfect geometries will be found to be poor representations of the electrodes used routinely. Thus it would be of importance to decide if (even approximate) solutions can be proposed to account for non-symmetric geometries, without the requirement of time and memory consuming sophisticated algorithms. In the mind of this author, one way to approach this problem would be to look for a geometry factor, g, characterizing the electrode. This "g" factor is defined by fitting the experimental behavior of the electrode by a diffusion equation such as:

$$\frac{\partial C}{\partial t} = D \left(\frac{\partial^2 C}{\partial r^2} + \frac{g}{r} \frac{\partial C}{\partial r} \right)$$

to a hypothetical electrode. For example, the planar electrode may have $g = 0$, the spherical or disk electrode may have $g = 2$, whereas the band or the cylinder electrode may have a $g = 1$. The validity and the operational effectiveness of such an approach is still an open problem. If such an approach or an analogous one were proved to be true, it would result in an easier treatment of data obtained at ultramicroelectrodes.

James Anderson contributed the following views on new opportunities that will be possible with the use of ultramicroelectrodes.

Ultramicroelectrodes and their cousins, including arrays of microelectrodes and dispersions of microparticles, enable a wide variety of experiments to be conducted that were previously impractical with macroscopic or monolithic solid electrodes. (The term *microelectrodes* will be used generically here to include all of the above mentioned electrode systems, with at least one dimension of very small size, unless specifically referring to aspects of particular types of systems.) The Workshop on Ultramicroelectrodes provided a very stimulating opportunity to participate in the discussion of a wide range of these new possibilities.

Summarized below are some of the key areas in which microelectrodes in general offer new opportunities, and some of the accompanying problems that must be solved to achieve optimum benefits. Also discussed are some special benefits which can be derived from the use of arrays of microelectrodes, which we are pursuing in our research group.

New opportunities made possible by microelectrodes

Electrochemistry in "unusual" places (relative to conventional practice):
- *In vivo*, in cells, nerves, tissue, etc.
- Liquids of low conductivity, which contain little or no electrolyte
- Gas phase
- Supercritical fluids
- Solid state (frozen solution, glasses, other solids)
- Evaluation of microstructure and/or morphology of heterogeneous or spatially nonisotropic media—2- or 3-dimensional

Microsensors

- Small, portable monitors for remote monitoring with minimal power requirements, possibly in hostile environments
- Disposable instruments on a chip
- Enhanced analytical sensitivity, detection limits, and/or selectivity, using arrays of sensors, with or without selective coatings and/or patterns of applied potential

Investigation of very rapid events

- kinetics - heterogeneous and homogeneous
- reaction mechanisms
- new media previously inaccessible

Monitoring processes at the molecular level

- Stochastic phenomena - random motion of small numbers of molecules, Poisson statistics
- Single events/single electron counting (still in early stages)
- Very small electrodes, or very short times

These studies are only a few of the possibilities opened up by the introduction of microelectrodes in electrochemical research. Undoubtedly, many more possibilities will open up as the field grows.

New problems to solve

Three general problem areas stand out: the mathematical description of phenomena frequently avoided with macroscopic electrodes; the physical and mathematical interpretation of new phenomena previously not observable with macroscopic electrodes; and significant materials characterization and the fabrication problems involved in making useful, robust, and reliable devices.

Mathematical and physical interpretation problems

- Nonuniform access problems - complex mathematics of mass transport are required due to geometric constraints

- Coupled modes of mass transport - diffusion, migration, and convection - frequently, two or three different modes are simultaneously important, unlike macroscopic electrodes, where greater control is feasible, but necessary experiments are not

- Coupling of Faradaic processes with diffuse double-layer relaxation (as well as analyte migration) in very dilute electrolyte with small microelectrodes, where the diffusion field may exist entirely within the confines of the diffuse double-layer

- Modeling of effects of solution and electrode resistance and current distribution effects on response at devices with one or more very small dimensions

- Statistical modeling of events where Gaussian statistics (conventional diffusion/Fick's Laws) gives way to Poisson statistics (small numbers of events), and single-event phenomena, where reaction media cannot be treated as continuous, but rather as discrete

Materials characterization and fabrication problems

Microelectrodes present severe problems of fabrication and characterization, arising from the importance of perimeter/area ratio as feature size decreases. Problems include:

- Materials compatibility - especially sealing and adhesion

- Susceptibility to poisoning due to small active area

- Resistivity problems

- Reliable methods of controlling and/or renewing the electrode surface - geometry, possible blockage by impurities or electrode reaction products, etc. - whether by polishing, fracturing, annealing, or other techniques

- Reliable construction of arrays of microelectrodes with integrity assured for all electrodes

- Robustness and reliability of fabricated devices
- Improved methods of fabrication for low cost, reproducible devices with a wide range of electrode materials and insulating sheaths, substrates, or supports, compatible with a wide range of media
- Methods for fabrication of integrated microsensors with signal processing and amplification incorporated into the device

Microelectrode array possibilities

Significant advantages can be achieved by the use of microelectrode arrays, whether all electrodes are held at a common potential, or of electrodes held at more than one potential. It can readily be demonstrated that an array of electrodes exhibits a cooperativity whereby the signal/noise ratio can be enhanced, relative to one or more solid electrodes occupying the same overall geometric dimensions. This cooperativity arises because arrays can exploit motions in more orthogonal dimensions than can be used effectively at monolithic macroscopic electrodes, whether working in stationary solution or in a flow stream. These orthogonal motions are accessible because the boundary conditions at adjacent electrodes of an array differ from one another, introducing an anisotropy into the mass transfer field.

The response per unit area increases with the dimensionality of mass transfer, so that arrays inherently have higher sensitivity per unit of active area than macroscopic electrodes. In addition, judicious placement of arrays of microelectrodes at two or more different applied potentials enables experiments to be carried out which are practically inaccessible with macroscopic electrodes. For example, shielding, deshielding, recycling amplification, and simultaneous voltammetric current-potential response monitoring is feasible with arrays of microelectrodes, depending on the geometry and applied potential combinations chosen. Such experiments are frequently feasible in flow streams or under steady state conditions which require no complex apparatus or potential programming capabilities, yet information can be acquired comparable to that obtained from relaxation experiments (e.g., potential step, etc.) or rotating electrode experiments (e.g., rotating ring-disk).

One of the most important areas in which arrays can be used to advantage is in steady state analysis. In a flow stream, constant-potential amperometric detectors generally achieve better signal/noise ratios and detection limits than potential-programmed detectors, principally because of the difficulty of discriminating against charging current and background current frequently associated with complex changes in potential. Use of arrays of microelectrodes enables experimentation at constant potential, for optimum signal/noise ratio and background discrimination, while retaining the qualitative identification and discrimination capabilities of the potential-programmed techniques.

We have been modeling and experimenting with arrays as amperometric detectors in thin-layer channels under conditions of laminar flow, using the backward implicit finite difference method. We have shown that significant enhancements in signal/noise ratio are to be expected by using arrays of microelectrodes, whether at a single applied potential or at multiple applied potentials.

We have presented theory for an array of parallel strip microelectrodes held at a single potential, and oriented perpendicular to flow on one wall of a thin flow channel. The results predict the feasibility of significant enhancement of signal/noise ratio relative to a monolithic macroscopic electrode of the same geometric area, due to convective

replenishment of the diffusion layer at inactive gaps between
active elements of the array (1). It is estimated that an order
of magnitude improvement in detection limits is achievable with
experimentally attainable microelectrode arrays. It was shown
that a random array of microelectrode sizes and spacing would
give response similar to that for an orderly array of uniform
size and spacing (1). It was also shown that the optimum
signal/noise ratio is obtained for a very sparse array
consisting of a very large number of very small microelectrodes,
although the lower limit of feature size and upper limit of the
number of microelectrodes will probably be set by problems of
excessive current density/electron transfer kinetic limitations
and resistivity of electrode materials. Sparseness is the most
important variable for enhancing signal/noise ratio, because
increasing the ratio of inactive area to active area increases
the degree of convective replenishment of the diffusion layer
and enables more efficient use of each active element, while
minimizing noise, many components of which are directly
proportional to active area. The number of microelectrodes has
a much weaker impact than the ratio of inactive area to active
area, which should be maximized.

The effect of diffusion parallel to flow was considered,
relative to convection, and microelectrode dimensions were
estimated at which edge diffusion must be considered. Edge
diffusion is important only at very small electrode dimensions
from channel thickness, and at very slow flow rates, otherwise
convection parallel to flow dominates. This differs
significantly from stationary electrodes in quiet solution,
where edge diffusion is prominent over a wide range of time
scales and electrode dimensions (1).

Experiments on the Kelgraf electrode, which can be portrayed
as a random array of microelectrodes, gave responses as a

function of the sparseness of the array which were in reasonable agreement with theoretical expectations for an orderly array with the same number of microelectrodes and same degree of sparseness (1). Significant enhancements of signal/noise ratio were observed relative to a solid electrode of comparable geometric area, consistent with predictions, although experimental enhancements were slightly lower than predicted.

We have developed the theory for an interdigitated array of strip microelectrodes of varying size and separated by variable gaps on one wall of a thin-layer flow cell, and held at alternating potentials (2). The optimum geometry for an interdigitated array was shown to consist of a large number of very narrow pairs of strips separated by the smallest feasible gaps. In this geometry, each microelectrode after the first serves as both a generator of product, and a detector of a species generated at prior electrodes. The collection efficiency of the array in a flow stream can exceed 70% for an array having more than 50 pairs of microelectrodes with minimal gaps between them, whereas the maximum feasible collection efficiency is ca. 40% for a single pair of electrodes. The relative advantage of the array increases as the relative dimensions of gaps relative to microelectrodes decrease, because each microelectrode helps to replenish the diffusion layer for its downstream neighbor, and increasing gap lengths allow increasing fractions of the reaction product to escape.

It has been shown for the interdigitated array that signal increases approximately with the number of pairs of alternating microelectrodes raised to the 1.1 power (3), and that the signal/noise ratio should increase as the number of microelectrode pairs raised to the ca. 0.43 power. Thus, it is predicted that the signal/noise ratio for an array consisting of 200 pairs of electrodes should be one order of magnitude better than for a single pair of macroscopic electrodes occupying the same total geometric area as the array. It should in principle be feasible to lower the detection limits ten-fold relative to conventional detectors, without any changes in electrochemical equipment other than the detector. This is a major advantage, as electrochemical detectors are already among the most sensitive available for flow analysis.

The applicability of an interdigitated microelectrode array has also been investigated theoretically for electrode kinetic studies (4), as a steady state analog in a flow channel of the rotating ring-disk electrode experiment. We have investigated the effect of electrode dimensions and number of microelectrodes on the electrode kinetics for an ec-type reaction involving a chemical step following electron transfer. Decreasing the microelectrode size and increasing the number of microelectrodes enables the investigation of significantly faster followup chemical reaction kinetics than feasible with a pair of macroscopic electrodes in series. Experimental evaluation of the theoretical predictions is underway for some reactions exhibiting rapid followup chemical steps. The use of asymmetric generator-detector pairs enables improved discrimination between simple, reversible, diffusion-controlled reactions and reactions undergoing chemical steps after electron transfer (4). An experimental program to verify this theory is underway (5).

We have also shown both theoretically and experimentally that the response of the microelectrode array at steady state in a flow channel is surprisingly immune to wide variations in microelectrode dimensions or gap spacings, at constant total geometric area and active area (6). The optimum array geometry consists of a set of microelectrodes of uniform length parallel to flow, separated by gaps of uniform length. However, simulated response was never diminished by more than 13% for a wide variety of asymmetric progressions of gap spacings or electrode lengths. Experimental results were in excellent agreement with theory for several microlithographically fabricated array geometries (6). This result will greatly simplify design of microelectrode arrays for use in flow channels, and gives assurance that imperfections in geometry of practical fabricated array devices will not give rise to excessive deviations from theoretical predictions.

A series of experiments has been carried out, and additional experiments are currently underway with a series of microelectrode arrays fabricated microlithographically, to test theoretical predictions and to assess the robustness and reliability of arrays fabricated by various methods. Experimental results for microelectrode arrays held at a single potential are in excellent agreement with theory for microelectrode arrays fabricated from gold and platinum on thermally oxidized silicon wafers and glass, respectively (7).

The following observations were made on microelectrode arrays fabricated on silicon wafers which had been thermally oxidized to form a 400nm thick layer of insulating silicon dioxide prior to metallization. The wafers were masked using a ca. 1.1mm thick layer of image reversal photoresist, which was developed to expose silicon dioxide where metallization was desired, and metallized by vacuum evaporation of the desired metal or metals. The final arrays were obtained by solvent stripping the remaining photoresist and the overlayer of metal adhering to it, leaving bare silicon dioxide or metal microelectrodes exposed (7).

When gold was deposited directly on silicon dioxide, adhesion was poor, as expected, but experimental response during the first several hours of use was in excellent agreement with theory for a wide range of geometries (7). The absolute currents measured generally agreed within 1-3% with theoretical predictions at all flow rates investigated, without use of any adjustable parameters. In addition, the dependence of current on flow rate rigorously followed the cube root dependence predicted under laminar flow conditions, and the signal/noise ratios for arrays relative to a monolithic electrode of the same total area, fabricated by the same process, were in good agreement with theoretical predictions. Approximately half an order of magnitude improvement in signal/noise ratio was achieved with arrays that were not yet of optimum geometry to achieve the tenfold improvement factor which our calculations predict is achievable with somewhat sparser arrays.

Response decayed by ca. 10-15% after several hours for the gold-coated silica arrays, and the dependence of current on flow rate degraded to a flow rate exponent of ca. 0.29, but a prolonged period of more than one day ensued, during which the response was nearly constant. Experiments with arrays fabricated with a thick underlayer of chromium metal under the gold showed substantially greater stability, with currents remaining in good agreement with theory for several days before decaying (8). However, the potential range was restricted due to enhanced background currents in the potential ranges more negative than ca. 0.3V and more positive than ca. 0.6V, which appeared to involve reduction and oxidation of chromium species. The reduction product was soluble in pH 7 phosphate buffer, and could be detected at downstream electrodes in the interdigitated array.

Enhancements of background current due to a ca. 600nm thick layer of chromium corresponded to relatively modest current densities of ca. $600\mu A/cm^2$ at the exposed edges of the chromium, but sufficient to vitiate use as interdigitated arrays. Additional studies are underway to develop a better adhesion layer (8).

The behavior of both the gold-only and the chromium-undercoated gold arrays is consistent with either fouling or development of microscopic blockage of sites on the microelectrodes, of sufficiently small size that diffusion parallel to flow must be considered. The latter interpretation is supported by microscopic examination of an array with a chromium underlayer after exposure to phosphate buffer at pH 7 for a week. Significant localized erosion of both the silicon dioxide and the metallization were evident in zones exposed to the flowing stream (8). We are presently investigating the sources of the observed behavior.

Future directions that seem especially promising include:

- New methods of fabrication of arrays that are robust and inexpensive

- Further investigations of composite, random arrays in light of their approximate adherence to theory for regular arrays and the wide range of robust devices feasible

- Shielding and deshielding experiments for analyte selectivity and identification

- Multipotential arrays for simultaneous identification and measurement of the potential-dependent response of species in flow streams

- Use of the principles governing microelectrode array behavior as a means to investigate the properties of composite materials by using them as microelectrode arrays to assess feature size and distribution

- Spatially-resolved experimentation, in which arrays of microelectrodes can be used to investigate the structure of microheterogeneous materials, including biological cells and organisms which take up or release redox components from or to their surrounding environment.

- Applications in trace analysis, e.g., in body fluids, where very low concentrations frequently are encountered

- Characterization of reactions and species under conditions where potential-programmed techniques are at a disadvantage, e.g., where currents due to surface reactions such as oxide formation interfere with currents due to reactions of interest.

The items outlined above are merely a subset of the range of possibilities accessible with microelectrodes, arrays thereof, or dispersions. Many exciting developments seem highly likely to emerge as the field develops. The field is very lively now, and significant advances seem likely to continue emerging for some time to come.

References

1. S. Moldoveanu and J.L. Anderson, J. Electroanal. Chem., 185 (1985) 239.
2. J.L. Anderson, K.K. Whiten, J.D. Brewster, T.Y. Ou and W.K. Monidez, Anal. Chem., 57 (1985) 1366.
3. J.L. Anderson, T.Y. Ou and S. Moldoveanu, J. Electroanal. Chem., 196 (1985) 213.
4. J.L. Anderson, T.Y. Ou and S. Moldoveanu, to be submitted to J. Electroanal. Chem., (1987).
5. T.Y. Ou and J.L. Anderson, work in progress, 1987.
6. L.E. Fosdick and J.L. Anderson, Anal. Chem., 58 (1986) 2481.
7. L.E. Fosdick, J.L. Anderson, T.A. Baginski, and R. C. Jaeger, Anal. Chem., 58 (1986) 2750.
8. L.E. Fosdick, Ph.D. Dissertation, University of Georgia, April, 1987.

Koichi Aoki contributed a synopsis of some of the work he has completed on the ultramicrocylinder geometry. He pointed out the following:

Microcylinder Electrodes

A microcylinder electrode, e.g., a carbon fiber electrode 0.1≈2mm in length and ca. 10μm in diameter, one end of which is exposed to a test solution and the other end of which is shielded with a glass capillary for the support of the electrode, has the following advantages:

1- Currents are not so small even at extremely thin electrodes because the surface area of the electrode surface can be controlled by changing the electrode length.
2- Microcylinder electrodes can be inserted into living tissues and be maintained at a specific target location without destruction of tissue.
3- The integrity of the geometry of microcylinder electrodes can readily be maintained during fabrication because these electrodes generally do not change mechanically or chemically.
4- Since the diffusion-controlled current at microcylinders has a logarithmic time dependence, high sensitivity is maintained even for electrolyses at long times.
5- Mass transport to a microcylinder electrode is governed by the cylindrical diffusion equation, some solutions to which are known. It is shown that the electrode not only works as an analytical tool but also has the potential of being an excellent device for kinetic studies.

We have derived various approximate relations for the electrochemical response of these devices and summarize them here.

Chronoamperometry (1)

The total current, $I(\theta)$, is expressed by

$$I(\theta) = 2\pi nFC^*Db[(\pi\theta)^{-1/2} + 0.422 - 0.0675 \log(\theta) \pm 0.0058 \{\log(\theta) - 1.47\}^2] \quad [1]$$

where $\theta = Dt/a^2$, and $\pm$ denotes $+$ for $\log(\theta) > 1.47$ and $-$ for $\log(\theta) < 1.47$. Here n is the number of electrons transferred per mole of reactant, F the Faraday constant, C^* the bulk concentration, D the diffusion coefficient, t the electrolysis time, and b and a are the length and the radius, respectively, of the microcylinder electrode. For large values of θ, the total current has little dependence on the radius. Thus a loss of sensitivity will not result even at extremely thin electrodes.

Linear sweep voltammetry (2)

The peak current, I_p, is expressed by

$$I_p = 2\pi nFC^*Db \,(0.446p + 0.355p^{0.15}) \quad [2]$$

where $p = (nFva^2/RTD)^{1/2}$. For $p > 0.5$, a plot of I_p against $(v)^{1/2}$ is roughly linear with a nonzero intercept.

Chronopotentiometry (3)

Since a steady state current is not obtained at a microcylinder electrode, a transition in potential-time curves for chronopotentiometry necessarily occurs. The transition time, τ, is expressed by

$$\tau = \tau_{sand}[1+2.47(nFC^*Db/I)+(1.76(nFC^*Db/I)^{1.97}] \quad [3]$$

where τ_{sand} denotes the transition time for the Sand equation.

Normal pulse (NP) and differential pulse (DP) voltammetry (4,5)

For pulse voltammetry at solid electrodes, it is necessary to make the pulse interval, τ_1, much longer than the pulse width, δ. The conditions under which the voltammograms are undistorted are $\tau_1 > 5\delta$ in the NP mode and $\tau_1 > 2\delta$ in the DP mode. Hence, the limiting current for the NP curve is expressed by Equation [1] in which θ is replaced by $D\delta/a^2$. The peak current of the DP voltammogram, ΔI_p, is given by

$$\Delta I_p = \tanh[(nF/RT)\Delta E/4] \quad I(D\delta/a^2) \quad [4]$$

where ΔE denotes the pulse height for the DP mode.

Pulse voltammetric current-potential curves for electrode kinetics (6)

Nonlinear diffusion permits kinetic control of some electrode processes. A technique of evaluating kinetic parameters from irreversible current-potential curves measured in normal pulse voltammetry has been developed.

Microband Electrodes

A microband electrode, which exhibits mass transfer similar to that at a microcylinder electrode, retains advantages 1 and 4. This geometry provides reproducible electrode behavior at sufficiently long times (tens of seconds).

Chronoamperometry (7,8,9)

A chronoamperometric curve at the microband electrode with width w is expressed approximately by

$$I = nFC^*Db[w(\pi Dt)^{-1/2} + 0.97]$$
$$-1.10 \exp[-9.90/|\ln(12.37Dt/w^2)|] \qquad [5]$$

At long times, the current exhibits a logarithmic time dependence.

Linear sweep voltammetry (10)

An expression for the linear sweep voltammogram at sweep rate v is given by

$$I/(nFC^*Db) = \frac{0.97 - 1.1 \exp[-9.9/\ln(5(\zeta+1)p^{-2})]}{1+e^{-\zeta}} \qquad [6]$$

for $p < 0.1$, where $p = (nFvw^2/RTD)^{1/2}$. For any value of p, the peak current I_p is expressed by

$$I_p/(nFC^*Db) = 0.439p + 0.713p^{0.108} + \frac{0.614p}{1+10.9p^2} \qquad [7]$$

Electrode kinetics at microdisk electrodes (11)

It is possible to evaluate kinetic parameters of irreversible reactions from the steady state current-potential curves at microdisk electrodes. The modified log-plot is given by

$$E = E^* - 2.3[RT/(1-\alpha)nF]$$
$$\log[\{1-(I/I_d)(1+e^{-\zeta})\}^{1.11}/(I/I_d)] \qquad [8]$$

with

$$E^* = E^o - 2.3[RT/(1-\alpha)nF] \log[(\pi/4)k^o \, a/D] \qquad [9]$$

where a is the radius of the disk, I the current at any potential, I_d the diffusion controlled limiting current and $\zeta = (nF/RT)(E-E^o)$.

References

1. K. Aoki, K. Honda, K. Tokuda and H. Matsuda, J. Electroanal. Chem., **186** (1985) 79.
2. K. Aoki, H. Honda, K. Tokuda and H. Matsuda, J. Electroanal. Chem., **182** (1985) 267.
3. K. Aoki, H. Honda, K. Tokuda and H. Matsuda, J. Electroanal. Chem., **195** (1985) 51.
4. S. Sujaritvanichpong, K. Aoki, K. Tokuda and H. Matsuda, J. Electroanal. Chem., **198** (1986) 195.
5. S. Sujaritvanichpong, K. Aoki, K. Tokuda and H. Matsuda, J. Electroanal. Chem., **199** (1986) 271.
6. K. Aoki, K. Tokuda and H. Matsuda, J. Electroanal. Chem., **206** (1986) 47.
7. K. Aoki, K. Tokuda and H. Matsuda, J. Electrochem. Soc. Jap., **54** (1986) 1010.
8. K. Aoki, K. Tokuda and H. Matsuda, J. Electroanal. Chem., in press.
9. K. Aoki, K. Tokuda and H. Matsuda, J. Electroanal. Chem., submitted.
10. K. Aoki and K. Tokuda, J. Electroanal. Chem., submitted.
11. K. Aoki, K. Tokuda and H. Matsuda, J. Electroanal. Chem., submitted.

A. Bezegh and J. Janata contributed the following on the preparation of ultramicroelectrodes, and the measurement of impedance at ultramicroelectrode structures.

Preparation of ultramicroelectrodes

Mechanically stable gold or platinum ultramicroelectrodes on planar ceramic substrates cannot be prepared without first plating the surface with Cr or Ti(W) and their oxides. These metals diffuse along the grain boundaries of the substrate and it is only a matter of time until the Au or Pt surface is compromised. The solution to the problem of interference of these metals with the electrochemistry is to etch the surface with an H_2O_2/EDTA mixture followed by re-electroplating a few monolayers of the noble metal. Such an electrode will be stable (by ESCA analysis) for several hours at room temperature. The argument that the surface does not show "Cr or Ti electrochemistry" cannot be considered accurate because even traces of these metals (which may be "invisible" to electrochemical measurements) can affect the charge-transfer kinetics. Edge exposed ultramicroelectrodes are superior in this respect.

Temperature effects

A rigorous treatment of the diffusion-kinetic problem of electrodes in which one dissipates heat must involve consideration of nonisothermal conditions in order to be meaningful. The problem is difficult because one has to consider at least three different heat capacities and three different heat flux coefficients in three different spatial regimes: the electrode metal, insulator region, and solution.

Warburg impedance of ultramicroelectrodes

The first question is that of why we were interested in ultramicroelectrodes. Originally we were interested in measuring exchange current densities of reactions. In order not to perturb the interface by application of an external signal, we utilized fluctuation measurements. We emphasize that by this method we measure under real equilibrium conditions spontaneous stochastic fluctuations of the electrochemical cell voltage. We will not present here details of the theory of fluctuation measurements; recent results in this area have appeared in the Journal of the Electrochemical Society.

Fluctuation measurements are limited by several factors. An important one is the frequency range available; another is the signal to noise ratio. The measurements are improved by decreasing capacitances, including the double layer capacitance, and by increasing the charge transfer resistance; thus ultramicroelectrodes offer a clear advantage in such an analysis.

Essentially no concentration gradient is produced at these devices, which is the driving force of diffusion. The product of flux and the diffusion coefficient is then zero. Subsequently, one can obtain pure kinetic information in the steady state (Fleischmann and Pons in Chapter 1).

The theoretical background of this phenomenon is the fluctuation-dissipation theorem, which establishes a direct relationship between the measurable power of fluctuations and dissipative forces; in this case, the real parts of the impedances. The proportionality factor is four times the Boltzmann constant, times the absolute temperature: (4kT).

With these points in mind, we applied the usual textbook impedance methods, with a double layer capacitance, charge transfer resistance and the Warburg term in the early interfacial model. This model didn't fit the experimental observations and our attention turned to the effects of hemispherical diffusion — due to the small electrode size.

We were unable to find microelectrode-impedance expressions

in the literature. Impedance expressions for a rotating disk do not give useful values in the limiting case where the rotation speed is zero, making it necessary to derive an impedance expression for disk ultramicroelectrodes.

A good starting point is to use the time decay curves known from chronoamperometry. The Cottrell equation is well described for linear diffusion, and several approaches have recently been published concerning nonlinear diffusion. Fourier and, specifically, Laplace transformations provide the direct connections between the time and frequency domains. Transforming the Cottrell equation yields the known impedance model of the electrochemical interface and transforming the approximate equations derived by Aoki-Osteryoung, Szabo, and Shoup-Szabo yields the impedance expression for disk microelectrodes. (Fleischmann and Pons now have provided the exact time dependent solution.) Because these authors only have expressions for short and long times, we obtained two expressions for high and low frequencies. The differences in the two limiting cases are minor (Figure 1).

Rearranging the impedance expression, it becomes obvious that the equivalent circuit of the hemispherical diffusion is a Warburg element in parallel with a resistance. The latter impedance comes from the hemisphericity. All of these elements have a slight frequency-dependence, coming from the low and high frequency expressions. The impedance model of an interface is shown in Figure 2. A similar situation was studied by Gerisher 35 years ago in which he discussed spherical diffusion around spherical electrodes.

These impedances appear in a Cole-Cole plot (Figure 3) not as semicircles, but as quarter-circles. Their centers are on the 45° line. In addition, the curves which connect the points on the quarter-circles belonging to the same frequency lie on

IMPEDANCE DERIVATION

* PLANAR ELECTRODE

    Cottrell $-\mathcal{L}->$ Warburg

* MICROELECTRODE

    Aoki–
    Osteryoung– $\mathcal{L}->$ Warburg
    Shoup–Szabo   for microdisc

FOR HIGH FREQU:

$$\frac{Zn^2 F^2 C^* a\pi}{RT} = \frac{1}{D + a(Djw)^{1/2}}$$

FOR LOW FREQU:

$$\frac{Zn^2 F^2 C^* a 4}{RT} = \frac{1}{D + 2a(Djw)^{1/2}/\pi}$$

Figure 1. The high and low frequency response.

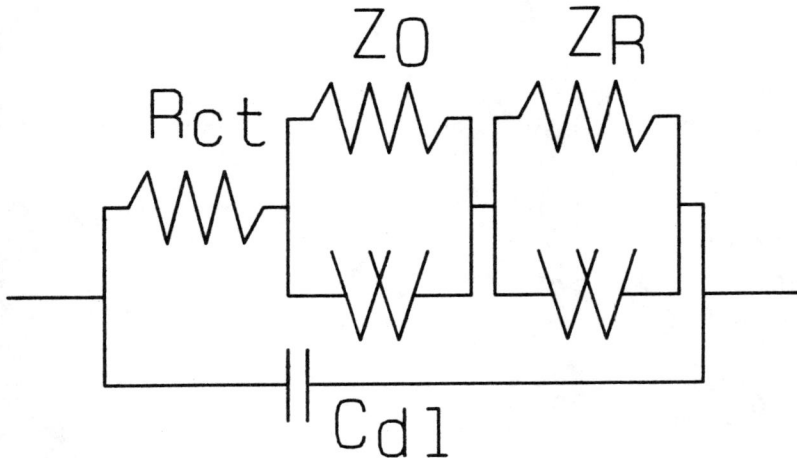

Figure 2. Equivalent circuit representation of the disk ultramicroelectrode.

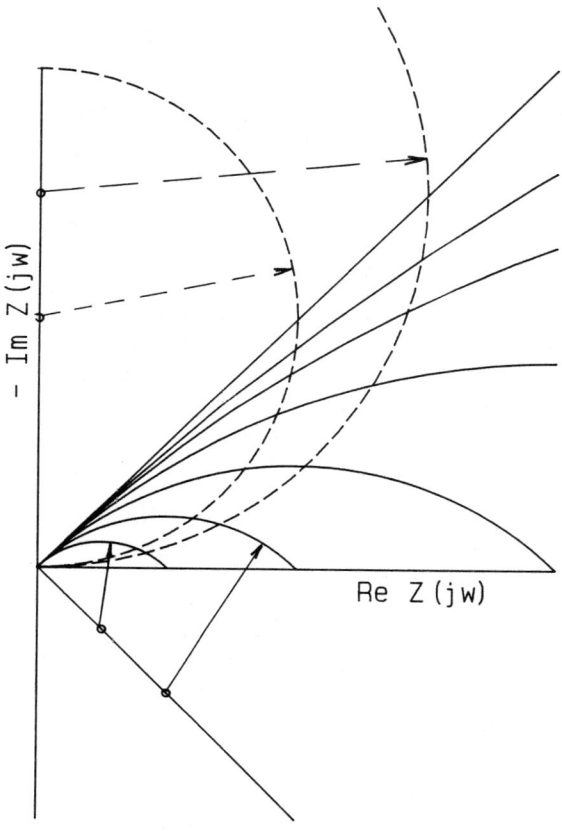

Figure 3. Cole-Cole plot for ultramicroelectrodes.

circles, with centers on the imaginary axis.

Further rearrangement and consideration of the impedance expression produced even more interesting results. The right-hand side denominator is a sum of the electrode area and of another term, which has an area dimension as well. This term is a product of the perimeter and a term designated as "radius increment". The entire denominator is the "effective diffusional area".

Using this idea a comparison can be made between a circular disk electrode and a ring electrode of the same geometrical area (Figure 4). The shaded areas are the geometrical electrode surfaces; the lined areas have the same widths at a certain frequency. They are the area increments.

Consider a ring microelectrode with the dimensions of radius 0.3cm and ring thickness 50nm which are the values of a published ring ultramicroelectrode (2). Plotting the results of the calculation on logarithmic scales shows how much larger the effective diffusional area is than the geometrical area (Figure 4). The vertical difference between the two curves is the "collection efficiency". The figure shows that this ring has a 100 times larger effective area than the corresponding disk electrode has in the usual frequency range. This approach is, however, a two-dimensional approximation. Using the effective diffusional area concept, several questions associated with microelectrodes and microelectrode arrays can easily be answered and if one the time domain answer is required, the "frequency-domain" can be transformed back.

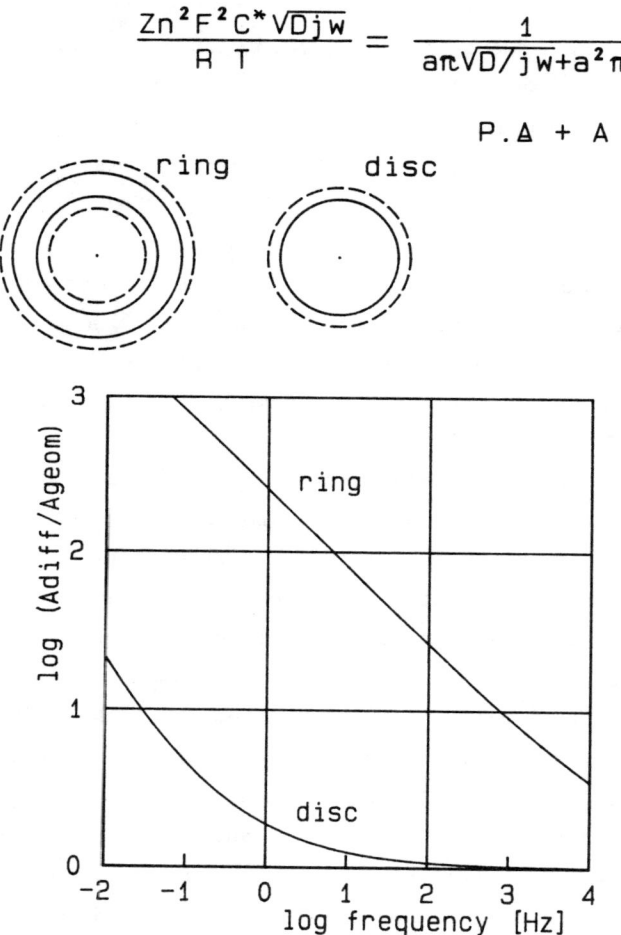

Figure 4. Comparison of the effective diffusional area for the disk and ring ultramicroelectrodes.

References

1. A. Bezegh and J. Janata, J. Electrochem. Soc., 133 (1986)
2. D. R. MacFarlane and D. K. Y. Wong, J. Electroanal. Chem., 185 (1985) 197.

John Bixler contributed the following on his work.

We have obtained some very interesting results in analytical experiments with flow jet cells (1,2). Electrochemical detection of sub-picomole quantities (femtoampere current levels) was shown to be quite feasible. We used 5μm carbon fiber and 50μm diameter platinum disk ultramicroelectrode mounted in a Metrohm wall jet flow detector cells to detect electroactive species. We were able to eliminate potentiostatic instrumentation in these experiments in favor of a low-noise, battery-driven two electrode configuration. It was shown that in flow detectors using ultramicroelectrode detection, little or no purposely added supporting electrolyte is needed.

References

1. J.W. Bixler and A.M. Bond, Anal. Chem., 58 (1986) 2859.
2. J.S. Bixler, A.M. Bond, P.A. Lay, W. Thormann, P. van den Bosch, M. Fleischmann and S. Pons, Anal. Chim. Acta, 187 (1986) 67.

Graham Cheek submitted a discussion of some of his work on microelectrodes and the study of organic ion associations.

Recent investigations of the electrochemistry of organic compounds at microelectrodes have demonstrated that much useful information may be obtained from studies which take advantage of the unique properties deriving from the very small dimensions of these electrodes. Of particular interest to the present study are investigations in nonaqueous solvents which ordinarily would not be possible at electrodes of the usual dimensions. Lines and Parker obtained very interesting results in benzene/tetrahexylammonium perchlorate (THAP) solutions at platinum electrodes of 30μm diameter (1), and recent work has shown that voltammetry in acetonitrile in the absence of added supporting electrolyte is feasible (2). Given the important role played by solvent properties in the course of organic reactions in general (3), the study of the electrochemical behavior of organic compounds in nonpolar media can provide a means to assess the importance of solvent-solute interactions in electrochemical reactions. Initial work has involved the electrochemical oxidation of hydroxyaromatic systems in benzene/THAP solutions. Infrared spectroscopy has also been employed to study the interactions of the compounds with the chosen medium.

Initial studies were directed toward an understanding of the oxidation of hydroquinone in benzene/THAP mixtures. Although hydroquinone oxidation has been studied extensively in more polar nonaqueous solvents (4), the use of microelectrodes (5μm diameter) allows investigation of this process in benzene provided that a suitable supporting electrolyte is added (1). Figure 1 presents a voltammogram for hydroquinone oxidation wave with a slope ($E_{3/4} - E_{1/4}$) of 70±1mV. This value was found to be appreciably higher at lower THAP concentrations (110mV at

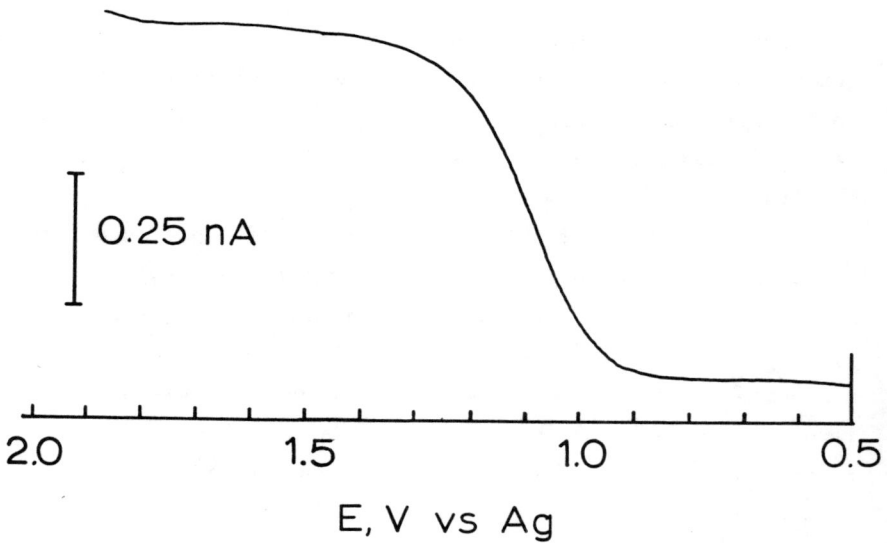

Figure 1. Voltammogram of 1.9mM hydroquinone in benzene/0.30M THAP. Working electrode: 5 micrometer diameter platinum. Scan rate: 30mV/s.

0.10M THAP, 82mV at 0.20M THAP) and remained constant at THAP concentrations of 0.30M to 0.50M, showing that effects due to high solution resistance are important at THAP concentrations below 0.30M. Voltammetry at scan rates up to 1000V/s produced a peak for the hydroquinone oxidation process but showed no peak for a corresponding reduction process on the return sweep; therefore, the oxidation does not produce a one-electron product which is stable on this time scale. Wave slopes were also obtained for hydroquinone oxidation in acetonitrile/TEAP (116mV at 0.10M TEAP, 90mV at 0.30M TEAP) and were seen to be roughly in agreement with those found above in benzene/THAP. These results are also in accord with a 95mV "peak width" reported by Eggins and Chambers from cyclic voltammetry of hydroquinone oxidation in acetonitrile/0.10M TEAP at a 1.3mm diameter platinum electrode, at which a two-electron process occurred (4). Although it seems likely that a similar pathway is followed in benzene/THAP, further studies are in progress to clarify the situation.

The effect of supporting electrolyte concentration on the hydroquinone oxidation process was studied in benzene, methylene chloride, and acetonitrile. It was found that, in acetonitrile/TEAP solutions, the limiting currents for hydroquinone oxidation decreased with increasing supporting electrolyte concentrations, due apparently to the increasing viscosity of the solvent system. As seen in Figure 2, however, the ratio of oxidation current for hydroquinone to that of ferrocene (added to the same solution) remains essentially constant over the range 0 to 0.50M TEAP. In the methylene chloride/TBAP system, a greatly different situation prevails as is evident from the continual decrease in this ratio observed as the TBAP concentration is increased. The effect is even greater for the benzene/THAP system, although it is not possible to determine

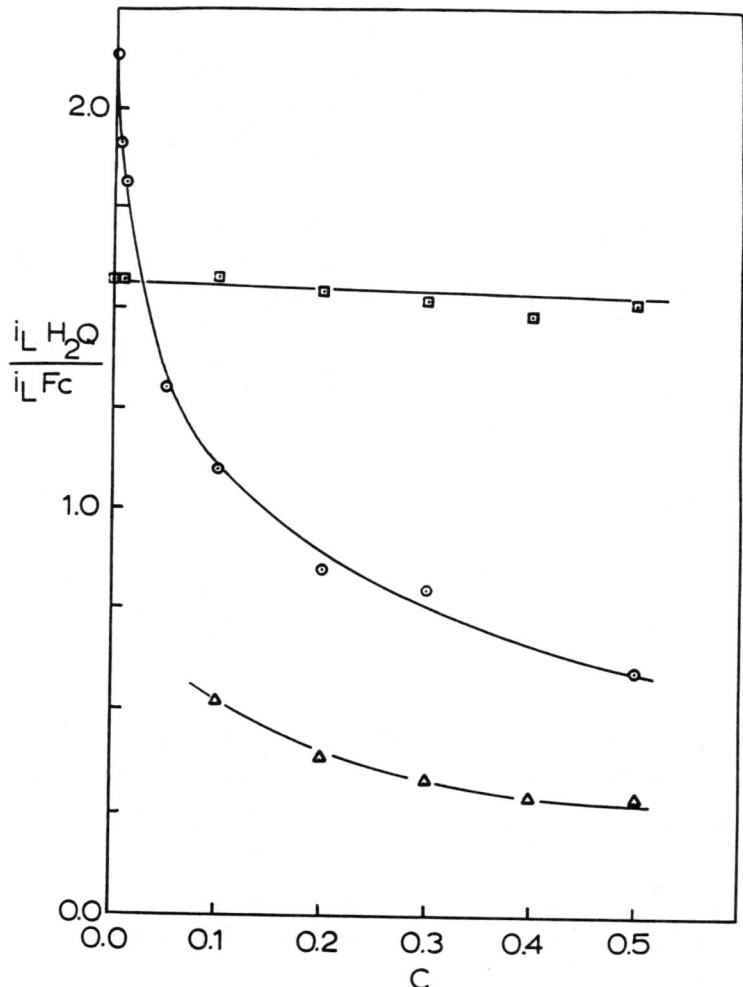

Figure 2. Ratios of limiting currents (hydroquinone/ferrocene) as a function of added supporting electrolyte.
□ Acetonitrile/TEAP: $[H_2Q]$ = 0.64 m$\underline{M}$ [Fc] = 0.59 m$\underline{M}$
○ Methylene/TBAP: $[H_2Q]$ = 0.91 m$\underline{M}$ [Fc] = 0.86 m$\underline{M}$
   chloride
▽ Benzene/THAP: $[H_2Q]$ = 1.90 m$\underline{M}$ [Fc] = 1.74 m$\underline{M}$

the ratio in benzene with no added supporting electrolyte.

Infrared spectroscopy of hydroquinone solutions in the media studied above helped to explain the behavior of the hydroquinone voltammetric responses. The region 4000-3300cm^{-1} was investigated in this work in order to study the effects of the media on the O-H stretching vibration for hydroquinone (Figure 3). In acetonitrile, a single broad band at 3425cm^{-1} was observed, the position of which was found to be independent of TEAP concentration. On the other hand, a single, rather narrow band at 3590cm^{-1} was seen for hydroquinone in methylene chloride, with another, broader, band appearing at 3470cm^{-1} as TBAF was added. An infrared spectrum of a hydroquinone/benzene solution (no added supporting electrolyte) showed two bands at 2564 and 3441cm^{-1} with approximately equal intensities. As THAP was added to the solution, the 3441cm^{-1} band shifted somewhat to 3411cm^{-1} and increased greatly in intensity (absorbance) as the band at 3564cm^{-1} decreased in intensity. These results indicate that hydroquinone in acetonitrile undergoes hydrogen bonding to the nitrile nitrogen so that the "free" O-H stretching band is not observed (5). The addition of TEAP apparently has no effect on this interaction. In methylene chloride solutions, the single band at 3590cm^{-1} is attributed to the free O-H vibration, and the shift to 3470cm^{-1} upon TBAP addition is taken as evidence of a hydrogen bonding interaction of the hydroxyl hydrogen with the supporting electrolyte (in particular, the perchlorate anion). The spectra for hydroquinone solutions in benzene indicate that significant self-association of hydroquinone occurs in the absence of THAP and that addition of THAP also results in an interaction with the supporting electrolyte similar to that seen in methylene chloride/TBAP solutions. These results correlate well with the polarities of the solvents involved. Interactions of hydroquinone with the

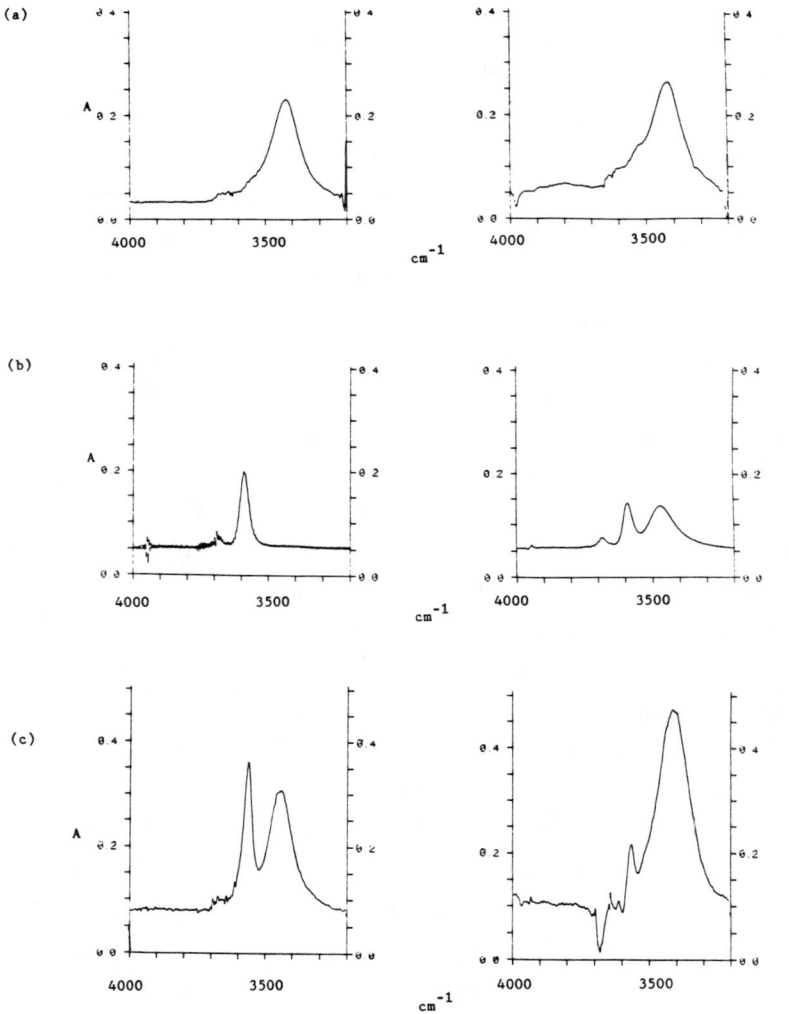

Figure 3. Infrared spectra of hydroquinone in various solvent systems. (a) 5.2mM hydroquinone in acetonitrile (left spectrum) 5.2mM hydroquinone in acetonitrile/0.10M TEAP (right spectrum)
(b) 4.0mM hydroquinone in methylene chloride (left spectrum)
   4.0mM hydroquinone in methylene chloride/0.10M TBAP (right spectrum)
(c) 10mM hydroquinone in benzene (left spectrum)
   10mM hydroquinone in benzene/0.15M THAP (right spectrum)

supporting electrolyte are expected to effectively lower the diffusion coefficient for hydroquinone as it is oxidized and to thereby produce the decreased current levels observed upon addition of supporting electrolyte.

In the preceding study, the functional group undergoing interaction with the supporting electrolyte is also oxidized in the course of voltammetric measurement of the effect. It was decided to include in this work a compound which possesses a hydroxyl function which is not directly involved in the electrochemical reaction of the molecule. The system chosen was 2-hydroxy-1,4-naphthoquinone, the reduction of which was studied in benzene/THAP solutions. As seen in Figure 4, an increase in THAP concentration results in a decrease of the current observed for quinone reduction (with respect to ferrocene oxidation). The effect seen for reduction of 1,4-naphthoquinone is much less than for 2-hydroxy-1,4-naphthoquinone; therefore, these results again indicate a specific interaction between the hydroxyl function and the supporting electrolyte.

In summary, electrochemical evidence of an interaction between hydroxyaromatic compounds and supporting electrolyte added to nonpolar solvents has been obtained. These findings are made possible by the use of microelectrodes in benzene/THAP solutions. Infrared spectroscopy of the solutions supports the conclusions of the electrochemical studies by showing a shift in hydroxyl stretching frequency upon addition of supporting electrolyte.

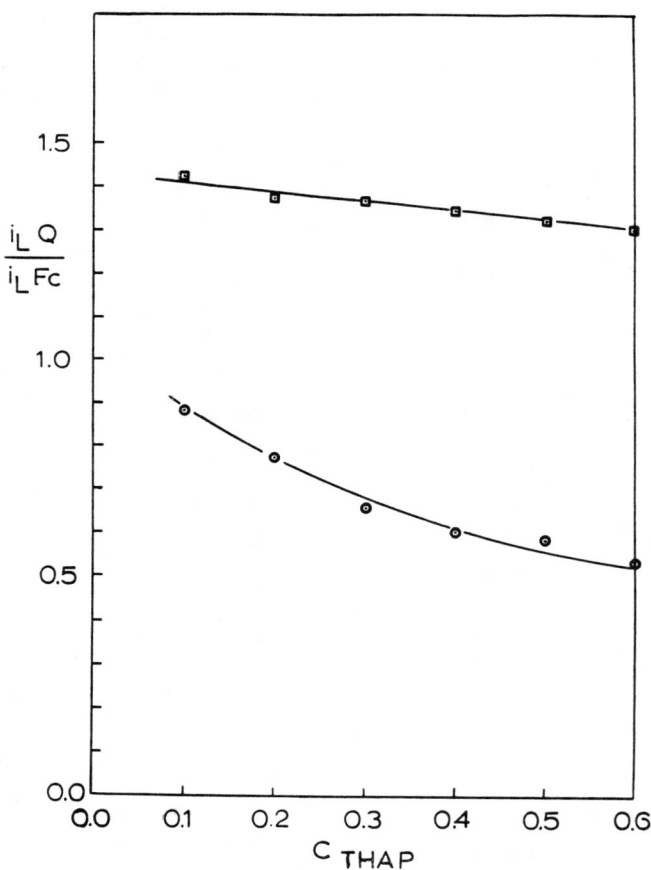

Figure 4. Ratios of limiting currents (quinone/ferrocene) as a function of added THAP concentration (benzene solutions).
□ 2.7m$\underline{M}$ 1,4-napthoquinone/2.0m$\underline{M}$ ferrocene
○ 2.7m$\underline{M}$ 2-hydroxy-1,4-napthoquinone/1.7m$\underline{M}$ ferrocene

References

1. R. Lines and V.D. Parker, Acta Chemica Scandinavica, <u>B 31</u> (1981) 369.
2. A.M. Bond, M. Fleischmann and J. Robinson J. Electroanal. Chem., <u>168</u> (1984) 299.
3. J. March "Advanced Organic Chemistry", Second Edition, McGraw-Hill: New York, 1977.
4. B.R. Eggins and J.Q. Chambers, J. Electrochem. Soc., <u>117</u> (1970) 186.
5. C. Laurence and B. Wojtkowiak Bull. Soc. Chim. France, (1971) 3124.

Andrew Ewing contributed the following material on very thin carbon ring ultramicroelectrodes.

Summary of Fabrication, Characterization and Application of Ultrasmall Carbon-Ring Electrodes

Fabrication and characterization by electron microscopy

The successful fabrication of ultrasmall ring-shaped carbon electrodes with total tip diameters as small as 1μm was presented at this meeting. These electrodes are fabricated by pyrolysis of methane inside quartz microcapillaries (1). The pyrolyzed carbon is deposited on the inner walls of the capillary. The tip is then filled with epoxy and cleaved to expose a ring-shaped carbon electrode. Scanning electron microscopy reveals that electrodes made in this fashion have total tip diameters as small as 1μm and ring diameters as small as 0.5μm. Ring thickness ranges from 50 to 100nm and the carbon formed appears to be fibrous as is expected for pyrolysis of hydrocarbons at moderate temperatures (approximately 1200°C (2)). The carbon in the final electrode is not porous to solvent, however, since the nonviscous epoxy used appears to fill in the spaces in the fibrous carbon and, when cured, forms a leak-proof barrier.

Characterization by voltammetry

Linear sweep voltammograms taken at moderate scan rates (approximately 100mV/s) are sigmoidal for ferricyanide (0.5$\underline{M}$ K_2SO_4; pH 3) reduction and dopamine oxidation (citrate-phosphate; pH 7.4) at carbon-ring microelectrodes. Voltammograms of dihydroxyphenylacetic acid in pH 7.4 citrate-phosphate buffer and ferricyanide at pH 8 are severely distorted

with no attainment of limiting current. This is apparently due
to slow charge transfer kinetics. In addition, anodization of
carbon-ring electrodes (rapid linear scans to 1.8V vs saturated
calomel electrode) alters the electrode response for voltammetry
of dopamine and dihydroxyphenylacetic acid at pH 7.4. After
treatment, the wave slope for voltammograms in dopamine
solutions becomes more nernstian, whereas that for
dihydroxyphenylacetic acid becomes more irreversible with an
accompanying shift of halfwave potential to more positive
potentials. This electrochemical treatment does not appear to
affect the voltammetry obtained in solutions of 4-methyl-
catechol. Identical pretreatment of disk-shaped carbon fiber
electrodes results in improved wave slopes for both dopamine and
dihydroxyphenylacetic acid, but again no effect on voltammetry
in solutions of 4-methylcatechol. This data might be
rationalized by the presence of different electron transfer
"sites" for oxidation of dopamine and dihydroxyphenylacetic acid
at carbon ring electrodes; these "sites" appear to be different
from that at carbon fiber electrodes.

Shielded carbon-ring microelectrodes

Ring-shaped carbon microelectrodes have been further
characterized by sealing them in large glass tubes (4mm in
diameter) with a glass-like epoxy (Epo-Tek 301). This procedure
permits resurfacing of the electrode by polishing and should, in
principle, permit comparison of experimental limiting currents
to those predicted by theory. Unfortunately, conventional
polishing of the electrode surface is extremely detrimental to
the electrochemical responses obtained at these electrodes.
Polishing with either diamond paste ($0.25 \mu m$) or alumina ($0.05 \mu m$)
on either a hard glass surface or a polishing cloth uniformly
results in no voltammetric response if the electrode is polished

to a mirror finish. Best results are obtained if a "rough" polish is implemented. This latter procedure involves polishing for approximately ten seconds at two stages beginning with 600 grit SiC paper and ending with 0.05µm alumina. When these polishing conditions are employed, the limiting current observed in dopamine solutions with shielded carbon-ring electrodes (average ring radius = 7 ± 1µm, N = 24) is only 22% of that theoretically predicted (3). In contrast, the limiting current at unshielded carbon-ring electrodes can be described in the following equation:

$$i_{lim} = (5.78 \pm 0.98) rnFDC) \quad (N = 6 \text{ experiments})$$

In this equation, r is the electrode radius, D is the diffusion coefficient, C is the bulk concentration of electrolyzed species and 5.78 is an experimentally determined proportionality constant. This constant is generally dependent upon electrode geometry and is equal to 4 for voltammograms at shielded disk-shaped electrodes when the voltammetry is obtained under conditions of steady state diffusion (4-7). However, this constant is also dependent upon shielding geometry (8) and is predicted to be equal to 5.2 when the radius of the disk-shaped microelectrode is half the total structural diameter. The limiting currents observed at cleaved and unshielded carbon ring electrodes agree surprisingly well with those theoretically predicted, whereas those at polished electrodes are consistently too small.

Electrochemistry at shielded carbon-ring electrodes appears to be highly dependent upon the current density during electrolysis. Wave slopes for the oxidation of dihydroxybenzylamine increase with the concentration of analyte (10mV/mM).

Intracellular voltammetry with carbon-ring microelectrodes

Structurally ultrasmall electrodes provide a unique opportunity to perform voltammetric measurements in environments too small to be accessible with previously available methods. We have successfully used carbon-ring electrodes to obtain voltammograms inside the giant dopamine neuron of the pond snail *Planorbis corneus*. The cell body of this neuron has a diameter of approximately 200μm (9,10) and is accessible following anesthetization of the snail with chloral hydrate and subsequent dissection. The inner volume of this cellular microenvironment is approximately 4nL. The tip dimension of the electrode employed must be small enough to allow the cell membrane to seal around the electrode after penetration into the cell. Empirical evidence suggests that optimum tip dimensions are a few micrometers or less and, consequently, electrodes with total tip diameter of 1 to 3μm are used.

Following implantation of electrodes in dopamine neurons, these cells are bathed with a dopamine solution and the subsequent change in intracellular dopamine concentration is monitored. Under these conditions, the rate of dopamine transport into the cell is rapid, occurring in 10-30s (peak) and this transport can be inhibited by the pharmacological agent amphetamine. These experiments provide the first evidence for a rapid uptake mechanism for dopamine into dopaminergic cell bodies.

References

1. Y.T. Kim, D.M. Scarnulus and A.G. Ewing, Anal. Chem., $\underline{58}$ (1986) 1782.
2. J.H. Je and J.-Y. Lee, Carbon, $\underline{22}$ (1984) 317.
3. A. Szabo, preprint (1986).
4. M.A. Dayton, J.C. Brown, K.J. Stutts, and R.M. Wightman, Anal. Chem., $\underline{52}$ (1980) 946.
5. J.O. Howell and R.M. Wightman, Anal. Chem., $\underline{56}$ (1984) 524.
6. M. Fleischmann, S. Bandyopadhyay and S. Pons, J. Phys. Chem., $\underline{89}$ (1985) 5537.
7. K. Aoki and J. Osteryoung, J. Electroanal. Chem., $\underline{160}$ (1984) 335.
8. D. Shoup and A. Szabo, J. Electroanal. Chem., $\underline{160}$ (1984) 27.
9. C. Marsden and G.A. Kerkut, Comp. Gen. Pharmac., $\underline{1}$ (1970) 101.
10. "The Characterised Dopamine Neuron in *Planorbis corneus*", in <u>Biochemistry of Characterised Neurons</u>, N.N. Osborne, ed., Pergamon Press: Oxford, 1978, pp. 81-115.

Larry R. Faulkner 225

Larry Faulkner contributed the following on ultramicroelectrode instrumentation.

Our group has developed three different new measurement systems for use with ultramicroelectrodes. All three were first reported at the Utah Ultramicroelectrode Workshop (1a). One of them is a simple intermediary for making possible the general use of commercial electrochemical instrumentation with ultramicroelectrodes. The remaining two systems implement new approaches to electrochemistry on very short timescales.

Current multiplication

One of the problems with ultramicroelectrodes is that the current levels generally are too low to be measured effectively with normal commercial electrochemical instrumentation. Individual investigators have solved this problem either by building special equipment themselves or by connecting the working electrode to a commercial picoammeter with a virtual-ground input, rather than to the working-electrode lead of a commercial potentiostat. These methods are effective, but have at least three drawbacks: (a) They have an ad hoc character that inhibits their widespread use by chemists who have no interest in instrumentation. (b) They can be rather expensive. (c) They normally defeat the computer interfaces that are available with the more sophisticated commercial instruments. This last drawback produces a significant loss of convenience in waveform generation, data acquisition, and data handling.

We have devised a very simple current amplifier that can allow any commercial potentiostat of which we are aware to carry out effective measurements with ultramicroelectrodes. We first reported it at the Utah meeting, but it has been subsequently described in print (1), so we will only outline its operational characteristics here.

Figure 1 is a circuit schematic of the current amplifier. The working electrode is connected to the summing junction of a current follower, which produces a voltage output proportional to the current at the ultramicroelectrode. This voltage is inverted by the second operational amplifier so that the sign convention for the current is ultimately maintained. The inverted voltage appears at the output of amplifier OPA 27 and is fed to the 10 kiloohm output resistor. The other end of this resistor, which is the output terminal of the packaged transducer, is connected to the working-electrode lead of a potentiostat. Normally, this is the summing junction of a current follower internal to the potentiostat. It is labelled CF in Figure 1. Since CF is a virtual ground, the inverted voltage at the output of OPA 27 is dropped fully across the 10 kiloohm output resistor, and the corresponding current is injected into the working-electrode lead of the potentiostat. In effect, the transducer simply receives the current from the ultramicroelectrode and transmits an amplified version of it to the potentiostat. The amplification factor is the ratio of the feedback resistance at OPA 104AM and the 10 kiloohm output resistance. In the device of Figure 1, amplification factors of 100, 1000, or 10000 are selectable by a switch. Thus, for example, a 100 pA current at the ultramicroelectrode can be transduced into a 1 microampere current, which is easily quantified by the potentiostat.

The transducer shown in Figure 1 can be constructed as an independent unit for a very low cost. It is used simply by interposing it between the working electrode and the working electrode's normal connection at the potentiostat. It normally will not interfere with any aspect of the potentiostat's operation. One only has to take care to rescale the currents reported by the commercial equipment in accord with the

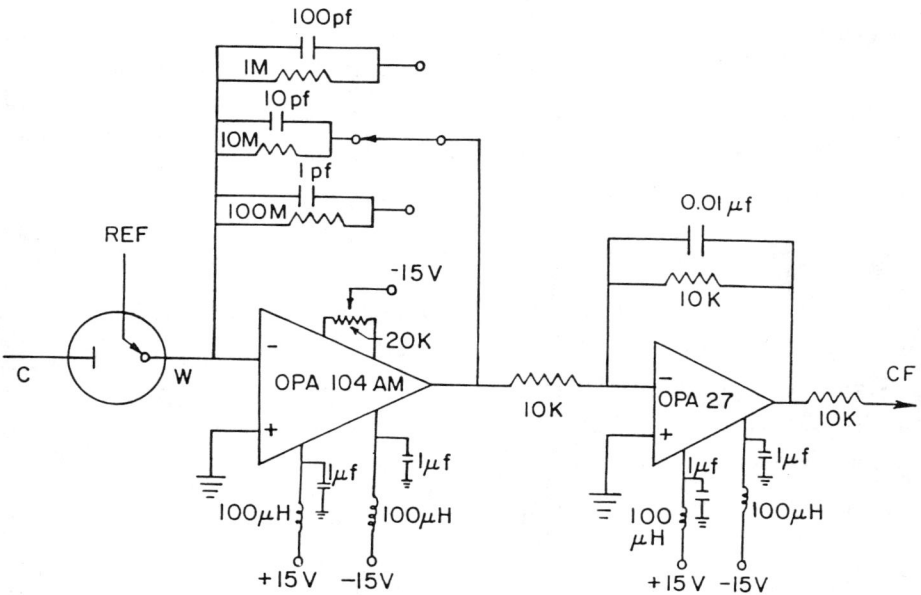

Figure 1. Circuit diagram of the current multiplier. Reproduced from Reference 1.

amplification factor used in the current amplifier. Figure 2 is an illustration of square-wave voltammograms taken with a BAS-100 cybernetic potentiostat [Bioanalytical Systems] at a 10 micrometer diameter carbon disk via the current transducer of Figure 1.

High-speed sampling via analog multiplication. One of the most important benefits of ultramicroelectrodes is access to much shorter timescales than are possible with conventional microelectrodes. As these timescales are explored, one can expect revelations of many mechanistic details that simply could not be time-resolved in earlier research. Realizing these benefits will require some new instrumental approaches, because equipment based on standard operational amplifier designs cannot be made to operate at timescales comparable to the shortest achievable cell time constants, which for the present seem to be in the range of 10ns. We have been exploring several ideas for using various kinds of intrinsically fast circuitry as elements in high-speed electrochemical instrumentation. One system that has proven effective is designed for Barker square-wave voltammetry at frequencies approaching 100 KHz. Its novel character arises from the use of an analog multiplier for high-speed demodulation and the use of conventional TTL for waveform synthesis.

In Figure 3 is an abbreviated schematic diagram of the apparatus, together with a set of timing charts showing how various events are interrelated. This equipment is still based to a considerable extent on operational amplifiers, but in the interest of effective high-speed operation, we eliminated both concatenated operational amplifiers and complex feedback loops. Waveform synthesis occurs at amplifier FG, and current transduction happens at CF. These both must have bandwidths considerably above 100 KHz. The waveform is a small-amplitude

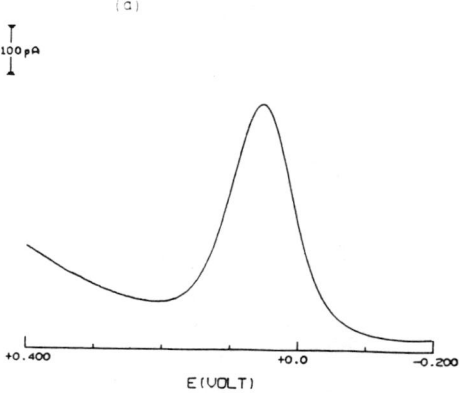

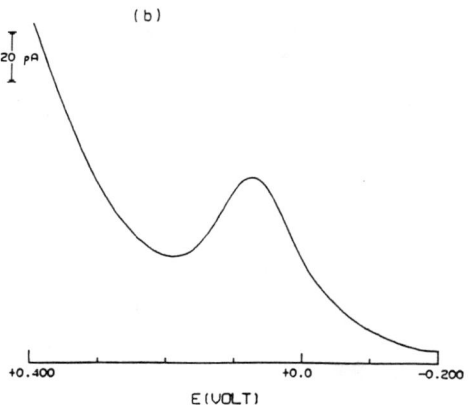

Figure 2. Osteryoung square-wave voltammograms for (a) 0.1 mM and (b) 0.01 M ferrocene in acetonitrile with 0.1 M tetrabutylammonium fluoroborate. Amplitude, 25 mV; step height, 4 mV; freq., 15 Hz. Reproduced from Reference 1.

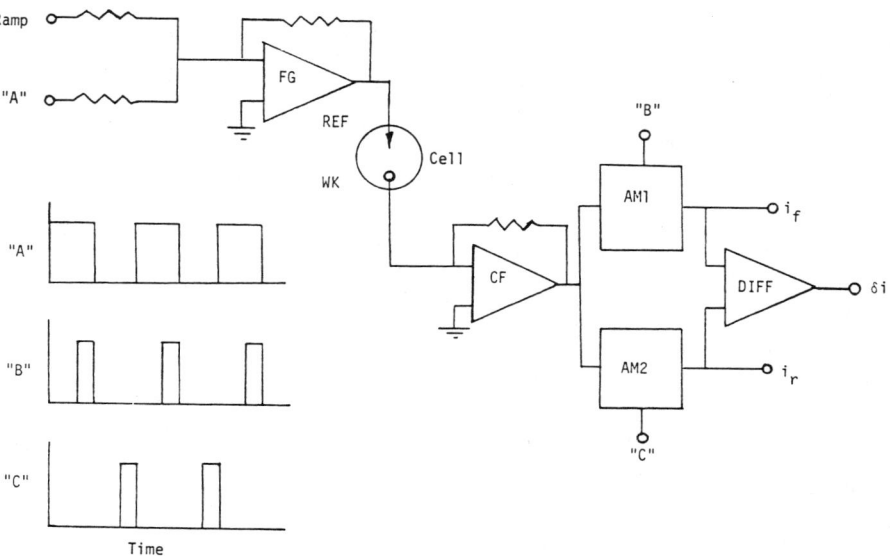

Figure 3. Instrument for square wave voltammetry in schematic form. AM1 and AM2 are AD534 analog multipliers. Logic circuitry (not shown) synthesizes the pulse trains at A, B, and C.

square wave superimposed on a slow ramp. This was achieved at
FG by summing a ramp (from an integrator circuit) with a
high-speed square wave that was simply a frequency-divided,
voltage-divided TTL pulse train delivered from a 555 oscillator.
The square wave component is illustrated in timing diagram "A."

One needs to be able to sample the transduced current late
in each half cycle. This is achieved in the two analog
multipliers AM1 and AM2 by feeding the TTL pulse trains at "B"
and "C" into the second inputs of the multipliers. These trains
were generated synchronously with that at "A" via monostable
chips. Waveforms "B" and "C" each cause the transduced current
waveform to be multiplied by a nonzero signal only over a period
late in one of the half-cycles, hence the output of each
multiplier reflects the transduced current only during the
desired sampling periods. At AM1, the output is a train of
interleaved current samples for the "forward" half cycles of the
square wave. At AM2, the output is a train of samples for the
"reverse" half cycles. These can be sent directly to a
recorder, which will filter out the high-frequency components
and provide forward- and reverse-current voltammograms.
Normally in Barker square wave voltammetry, one wishes to
examine the difference between forward and reverse current
samples, hence the outputs of AM1 and AM2 were also fed to a
difference amplifier DIFF, as shown in Figure 3. All of the
hardware downstream from the analog multiplier could operate on
a narrower bandwidth, because the rate of change of the
demodulated signals is low in square-wave voltammetry.

The base frequency used in our apparatus was 100 KHz.
Reliable measurements probably could be made with it at slightly
higher frequencies, but we were confined to 50 KHz by the time
constants of the Pt disk ultramicroelectrodes used in most of
our work. These electrodes were comparatively large, at 25

micrometer diameter, and they faced significant resistance in the acetonitrile working solutions that we normally employed.

Well behaved, bell-shaped square-wave voltammograms were obtained for various chemical systems. Figure 4 provides typical results. Standard heterogeneous rate constants on the order of 1 cm/s were evaluated by examining the frequency dependence of peak height. We have also explored the voltammograms of forward and reverse current samples as indicators of kinetic reversibility. A particularly striking result is shown in Figure 5, which is a plot of peak height vs. the square-root of frequency for ferrocene in acetonitrile. Even at 50 KHz, we were unable to detect a deviation from linearity, hence the standard rate constant for ferrocene appears to be greater than 1.5 cm/s. We expect to report more details in another forum soon.

Improvements in this system ought to make possible square-wave measurements at frequencies near 1 MHz, and conceivably as high as 10 MHz. The limitations will be imposed by the continued involvement of operational amplifiers. Analog multipliers, many of which were developed for FM radio communication, have quite high bandwidths and are very attractive for electrochemical demodulation in many contexts.

Current-to-photon conversion

Obtaining truly time-resolved electrochemical responses in the nanosecond regime is complicated by significant difficulties in the measurement of cell currents. One cannot reach cell time constants smaller than 100ns without using fairly small ultramicroelectrodes, hence one must accept low currents as a fact of life. Conventional techniques for measuring these currents would normally involve resistances larger than 100 kiloohms, and when these are coupled with stray capacitances in

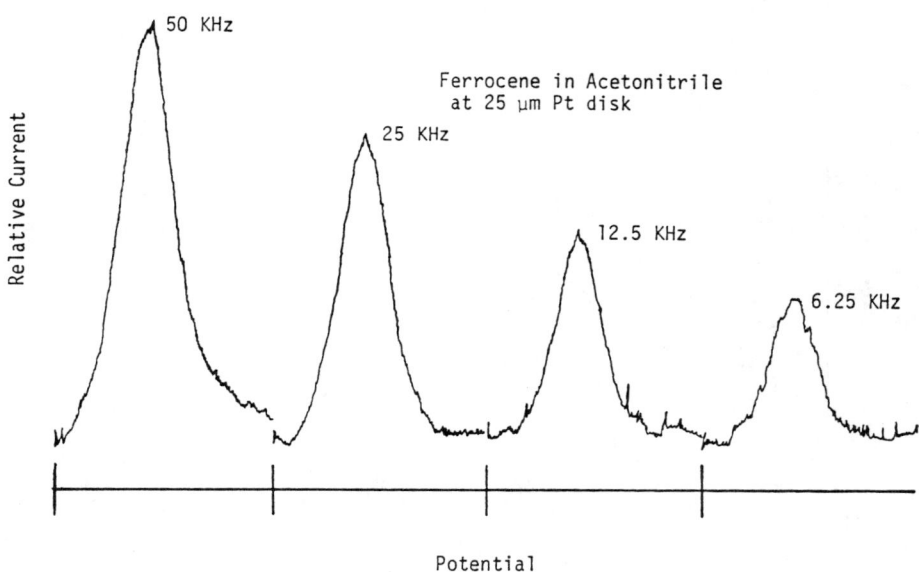

Figure 4. Barker square wave voltammograms for 0.1 mM ferrocene in acetonitrile containing 0.1 M tetrabutylammonium fluoborate. Distance between ticks is 600 mV. Curves are displaced horizontally for clarity.

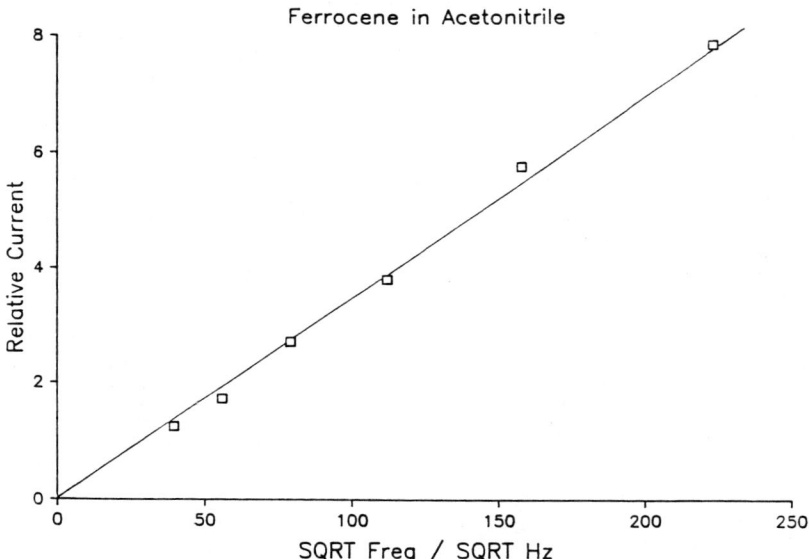

Figure 5. Relative peak current vs. square root of frequency for ferrocene. System as in Figure 4. Top frequency is 50 KHz.

the picofarad range, one finds time constants in the measurement circuitry that considerably exceed 1 microsecond. New strategies are needed for measuring low currents at high speed. An attractive approach is to convert the current to a photon flux, then to take advantage of the high speed photon detection circuitry that has evolved for time-resolved spectroscopy. We have investigated some prospects in this direction. Pons and Fleischmann discuss others elsewhere in this volume.

Electrons can be transduced to photons with fairly high efficiency by passing the electrons through a forward-biased light emitting diode (LED). At the time of the Utah workshop, we accomplished this conversion in the manner described in Figure 6. Potential steps were applied directly to the cell from a high-frequency waveform generator. The working electrode was held at virtual ground by the operational amplifier, which was configured like a current follower, but with the feedback resistor replaced by a pair of diodes. One of these was an LED, thus current flow in one direction only was transduced to light.

We elected to obtain time resolution via the method of time-correlated single photon counting. In this approach, one detects only a single photon for an average of 50-100 step experiments. To obtain good precision, many steps and step reversals are applied. Typically about 100,000 photons are detected, hence 5-10 million step experiments have to be performed. When each photon is detected, the delay time between the start of the corresponding step and the detection of the photon is recorded. Via a multichannel analyzer, the system builds up a histogram of delay times having the same shape as the time dependence of the light intensity, which in our experiment can be mapped into the cell current. The time-correlated single photon counting experiment is widely used to measure luminescence emission decays after delta-function

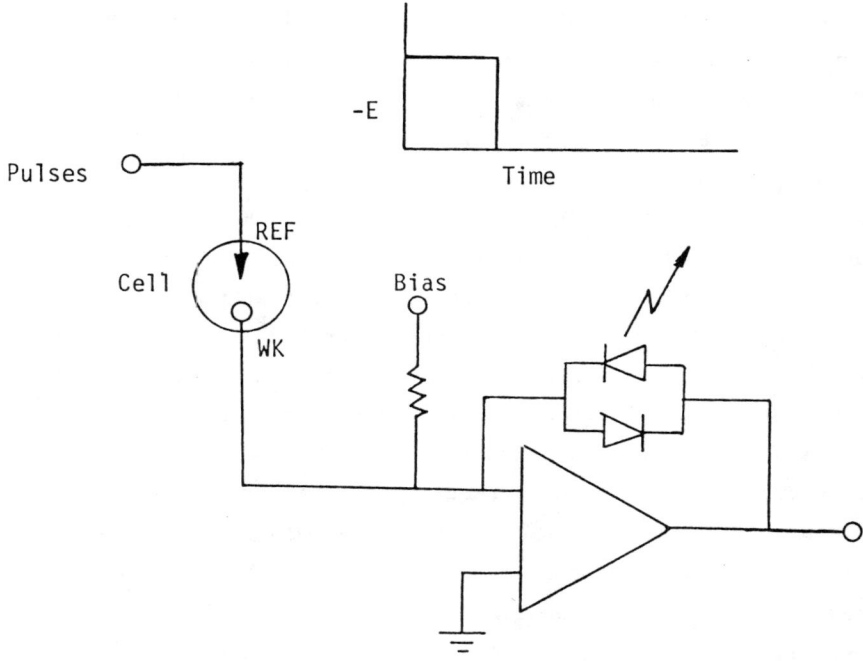

Figure 6. Schematic diagram of the current-to-photon converter. A bias is applied as indicated, so that the LED is emitting at a low level even at zero cell current. Pulses are applied from a fast pulse generator.

excitation of a sample. Present state of the art in time resolution is about 40 ps, so the method can handle anticipated electrochemical timescales with ease.

Figure 7 provides data that were displayed at the Utah workshop. One set is for a dummy cell of a series resistance-capacitance combination, while the other is for the formation of surface layers (oxide and adsorbed hydrogen) on platinum electrodes. These figures mainly establish the ability of the method to provide time-resolved response functions in the low microsecond time regime.

Since the time of the Utah workshop, we have considerably improved our techniques. We have found a means for measuring bidirectional currents, hence we can carry out double-step experiments. In addition, we have developed a method for calibrating the histograms, so that a current scale exists on the ordinate. Finally, we have performed kinetic studies of hydrogen atom adsorption and oxide film formation on clean platinum in perchloric acid. A manuscript describing this research is now in preparation.

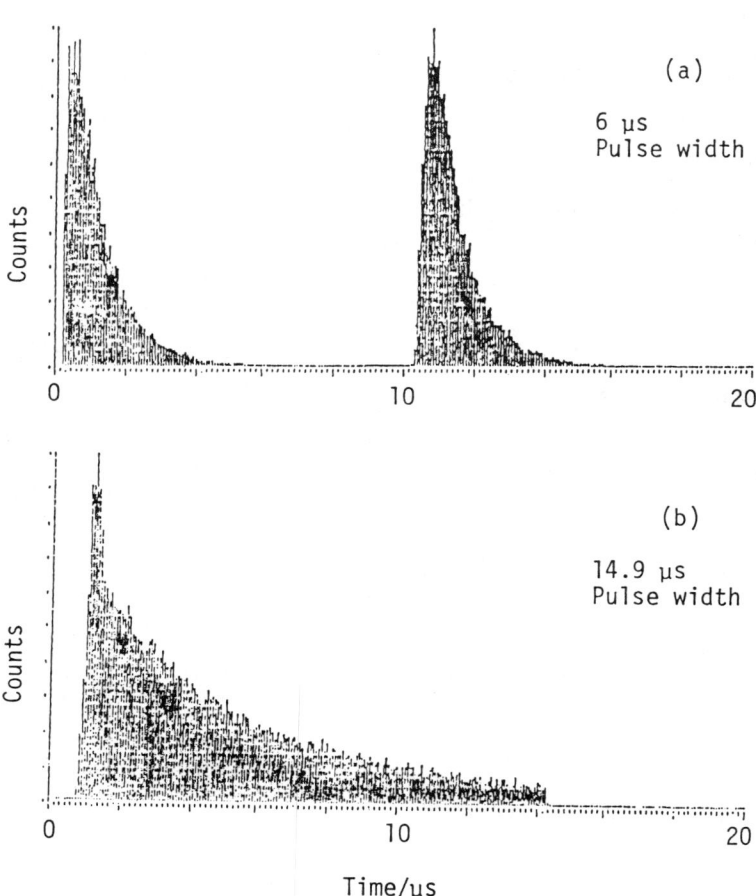

Figure 7. Data obtained with the apparatus of Figure 6 via time-correlated single-photon counting. (a) For an RC dummy with 1 microsecond time constant. Observed decay time was 1.2 microsecond. (b) Current transient in a cell with two Pt ultramicroelectrodes (25 micrometer diameter) in 0.2 M potassium sulfate.

Acknowledgements

We are grateful to the National Science Foundation for supporting this work under Grant CHE-86-07984.

References

1a. L.R. Faulkner, P. He, D. Ingersoll, H.-J. Huang and M.R. Walsh, communicated.
1. H. J. Huang, P. He, and L. R. Faulkner, Anal. Chem., 58, 2889 (1986).

Stephen Feldberg contributed the following regarding applications of and the theory pertaining to ultramicroelectrodes.

Applications of Microelectrodes

Biological sensors

This area is where the real applications have been thus far. The traditional "micro-salt bridge" has been a standard tool for neurophysiologists for the better part of a century. Metal or metal-like microelectrodes have been used for more than 20 years. Ralph Adams originally, and now, Mark Wightman have developed the use of microelectrodes into a sensitive and valuable tool. The critical point here is that there is at present no other way to do these *in situ* measurements.

Novel physical-chemical measurements

The work of Pons, Fleischmann et al. has demonstrated some unique capabilities of microelectrodes, e.g., electrochemistry in systems with no supporting electrolyte, even in gases and of species (e.g., some rare gases) that are not normally considered to be electroactive. Here, too, things are being done that cannot be done any other way. Some unique analytical sensors might be forthcoming.

Electroanalytical enhancement

Some ideas have emphasized the enhanced signal-to-noise ratio at a microelectrode or microelectrode array. Most (not all—see above) of these analyses can be done in other electrochemical or nonelectrochemical ways and will be bucking the usual preference for optical or other electromagnetic detection. This is not to say that some electroanalytical analysis would not be the optimum approach to a given problem. It is however, less likely to be the choice.

Theory

The mathematical models for most conventional geometries have been adequately analyzed. Once a geometrically well characterized microelectrode has been produced the appropriate theory can be developed: certainly by brute force finite difference methods; perhaps by some more efficient implicit finite difference or other numerical algorithms, and least likely by an analytic solution. Theory will not be the rate-determining step in this field.

Microelectrodes are most likely to be used where they are the *only* way of making a measurement. They are not likely to be used widely where they are only a slightly better way to make a measurement.

Whatever developments evolve, theory is not going to present an impediment.

Contributions

Martin Fleischmann presented the following lecture on connections of ultramicroelectrodes with other fields of research.

Connections with other fields of research

Recent applications of microelectrodes include electroanalytical measurements (using the chronoamperometric or chronopotentiometric responses), the measurement of the kinetics of fast electrode reactions (from the polarization waves) and of the kinetics of fast reactions in solution coupled to electrode reactions (using, for example, the radius dependence of the kinetically limited current). This work is reviewed elsewhere in this volume. These applications are dependent on two special properties due to the spherical (or quasi-spherical) concentration and potential fields surrounding the microelectrodes, namely:

(i) the rapid attainment of high rates of steady state diffusion

(ii) the low ohmic potential drops in the solution.

The second effect is of particular importance to other applications of microelectrodes as it allows measurements to be made on novel systems under unusual conditions such as in

a) nonpolar solvents in the presence of appropriate suppor electrolytes such as tetrahexylammonium perchlorate (1-3)

b) polar solvents (4-9) and mixtures of polar and nonpolar solvents (10) in the absence of support electrolytes

c) low temperature glasses and eutectics (11)

d) the gas phase (12).

Measurements of the type a), b), and c) have been made with conventional microdisk electrodes while b) has additionally been demonstrated using the bipolar electrolysis of dispersions of metal particles (8,9). It should be noted that ions are invariably generated (or consumed) in electrode reactions *viz.*

$$Ox + e \rightleftharpoons Rd^-$$

or

$$Ox^+ + e \rightleftharpoons Rd$$

so that conditions can be chosen to increase the conductance in the vicinity of the microelectrodes. Since the bulk of the ohmic losses take place in this region (the resistance is proportional to the resistivity and the inverse of the electrode radius), the ohmic losses are small and calculable (13) or measurable (10) in particular in the kinetically controlled region of the polarization curves. In measurements of the type d) the current will usually (but not necessarily) flow over the "insulator" surface separating the disk from the surrounding ring electrode while gas phase species diffuse to the disk (or, most likely, to the edges of the disk) through the quasi-spherical diffusion field. The conditions are therefore somewhat different to those for measurements of the type a)-c).

The investigations of these novel systems in turn prompt a series of questions about the connection between such electrochemical measurements and other established fields of research. For example we can list some possible connections in Table 1. The divisions are naturally somewhat arbitrary. For example, consideration of the conditions for field ion microscopy or high field ionization are equally relevant to studies in the liquid phase as to the gas phase. We note that research in the "other areas" listed in Table 1 usually has objectives (such as the imaging of surfaces, the generation of parent ions for mass spectroscopy, the charging of fuels or polymers) which do not include the study of the interfacial charge transfer steps. However, the total description of the processes must certainly include that of the surface reactions. It should prove possible

Table 1. Connections between the investigation of electrochemical systems using microelectrodes and other fields of research.

Measurements with microelectrodes in the	Relevant areas of research
liquid phase cf a), b)	high field charge injection scanning tunnelling microscopy heterogeneous catalysis
solid phase cf c)	charge transfer in solids reactions in solids charge injection into insulators
gas phase cf. d)	thermionic emission field emission microscopy scanning tunnelling microscopy high field ionization (generation parent ions high vacuum experiments heterogeneous catalysis

to design many interesting new experiments e.g., the examination of the structure of polarized surfaces using high vacuum techniques such as RHEED (high vacuum rather than ultra high vacuum). In this vein one can ask: would it not be possible to study electrochemical reactions induced by an electron beam (cf. e-beam lithography) which electron microscopists seek to avoid by coating surfaces with electronic conductors?

Connections with other areas of research in electrochemistry

The investigations using microelectrodes also prompt a series of questions regarding the connections with other areas of research in electrochemistry. Here we will consider three aspects:

      (i) novel ways of investigating established systems

      (ii) the reinterpretation of the behavior of known systems

      (iii) earlier investigations of the behavior of microelectrodes.

These divisions are naturally again somewhat arbitrary; we will illustrate (i)-(iii) with a small number of examples: the list could certainly be extended.

(i) Novel ways of investigating established systems

Some investigations of this type are already being carried out e.g., measurements on polymer modified electrodes (which can also be related to topic c) above). A new example would be the photoelectrochemistry of dispersions of semiconductors or of microemulsions subjected to bipolar electrolysis. These systems have hitherto been investigated at the mixture potentials e.g., see (14-17): the measurements with dispersion of metal particles (8,9) show that it should prove possible to derive much new information from the Faradaic behavior of such systems.

(ii) Reinterpretation of the behavior of known systems

An example of such an approach is the investigation of the behavior of electrodes showing a nonuniform activity over the surface, i.e., a variation of the rates of reaction with positions. We can include under this heading the effects of adsorbates on electron transfer reactions (especially for adsorption processes showing higher order phase transitions) as well as the effects of the distribution of catalyst sites on the rates of electrocatalytic reactions (for an extreme example see (iii) below).

In a different vein we note that it should be possible to make electrodes having at least one dimension comparable or smaller than the Debye screening length e.g., by using line or ring electrodes. Measurements with such electrodes should therefore provide a new tool for probing double layer structure.

(iii) Earlier investigations using microelectrodes

The term "earlier" here refers to investigations which, although they had somewhat different objectives to those of present research, actually embodied many of the concepts now being used.

One of the best known examples is the use of microelectrodes to stimulate and monitor physiological processes. Examples of direct interest to the study of electrochemical kinetics include the ion transduction processes in lipid bilayers (for review see (18)). For example, the application of an appropriate voltage difference to bilayers containing the polypeptide alamethicin induces a current-time series showing transitions between discrete current levels (see e.g., (19-22)). Transitions between levels are associated with the insertion/reorientation or removal/reorientation of single molecules of the peptide which aggregate to form pores reaching a maximum size of 4-5

molecules (for other references, see (22); analysis of the data
and modelling, see (22-24); modelling of other systems, see
(18)). It can be seen that in systems of this kind the
generation of the current transducing element defines a
"microelectrode" of molecular dimensions in a substrate having
an area typically $1mm^2$. The use of fire polished capillaries to
isolate small patches of membranes has greatly extended the
scope of such measurements e.g., see (25). It should also be
noted that in systems of this kind the subsequent essentially
deterministic transmembrane ion current amplifies the random
aggregation/removal of the pore forming molecules so that the
kinetics of these processes and the energetics of the states
become measurable at the molecular level (26) (see also comments
on nucleation below).

A second large group of phenomena which depend on the
formation of microelectrodes is that of the electrocrystallization of new phases. For large electrodes these processes
depend on the nucleation and subsequent growth of centers of the
new phase i.e., we are dealing with the growth of ensembles of
microelectrodes. In view of the very small dimensions of the
centers (which may well have radii 1-10nm) diffusion is rapid so
that the processes are controlled by the kinetics of the surface
reactions and ohmic potential losses in solution are very low.
Indeed one could argue that the successful operation of many
electrochemical devices depends on the formation of such
ensembles of microelectrodes (e.g., the oxidation of $PbSO_4$ to β-
PbO_2 in the lead acid battery (27)).

When electrocrystallization proceeds via the nucleation and
growth of two-dimensional centers of monomolecular height (such
as in the formation of a variety of anodic films e.g., (28-30)
or the deposition of silver on perfect single crystal substrates
(31)) the processes can be modelled as the growth of circular or

polygonized microelectrodes. The deposits are usually highly oriented so that we observe electrode reactions in defined crystallographic directions i.e., typically lattice formation at the edges of layer planes of the lattice. In these cases one characteristic length scale of these microelectrodes is of atomic dimensions and, in consequence, the observable rates of reaction are very high. The steps propagate over the surface into regions of the solution where the reagents have not been depleted so that, in contrast to stationary "line" electrodes very high rates of steady state mass transfer can be achieved (several hundred cm s^{-1} (30). The modelling of mass transfer to arrays of such edges (32) (such as those generated by emergent screw dislocations (33-39)) has in fact been known for quite some time.

Catalytic reactions at the edges of island films and at "holes" have been studied, e.g., hydrogen evolution at ruthenium layers (36) and the oxidation of CO adsorbed on Pt (37).

The application of microelectrodes, however, in turn opens up new opportunities for the study of phase growth as it becomes possible to investigate the formation of single (31,38,42,18) or, at the most, a few growth centers. We illustrate these processes with the growth of a single droplet of mercury (41), Figure 1, and of a single center of α-PbO$_2$, Figures 2 and 3, on carbon microelectrodes. The first process has usually been studied at higher overpotentials (40) where the onset of diffusion controlled growth ($i \alpha t^{1/2}$) defines the time of "birth" of the first nucleus. However, Figure 1 shows that the reaction is kinetically controlled in the initial stages and that it is now quite feasible to make measurements on such electrodes having characteristic dimensions of $\approx$ 10nm. This is shown more clearly by the second example, which is a much slower reaction, where the time of birth is marked by the onset of the

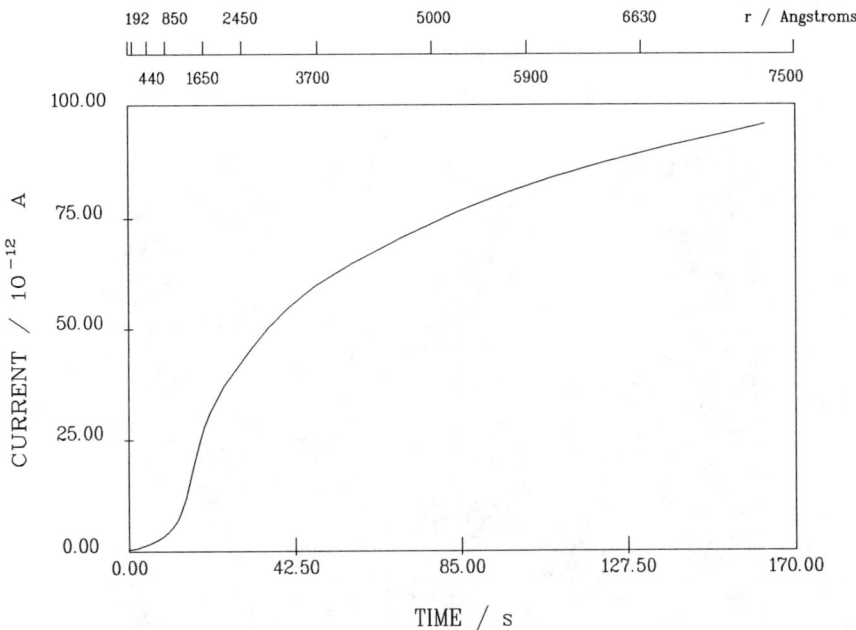

Figure 1. Current-time transient for the deposition of a single droplet of mercury from a solution 0.2m$\underline{M}$ Hg$_2$(NO$_3$)$_2$ + 10m$\underline{M}$ HNO$_3$; η = 5mV; 4μm radius carbon disk electrode.

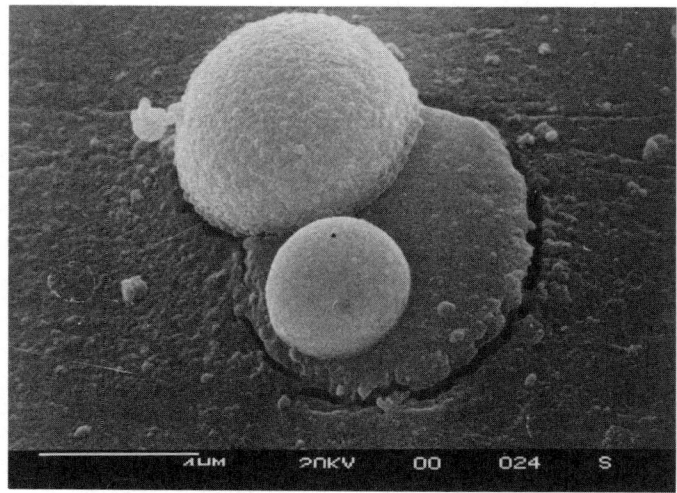

Figure 2. The deposition of two growth centers of α-PbO_2 onto an 8μm diameter C-microelectrode. Solution composition: 0.1$\underline{M}$ Pb(Ac)$_2$ + 1$\underline{M}$ HAc; η = 400mV; deposition time = 120s.

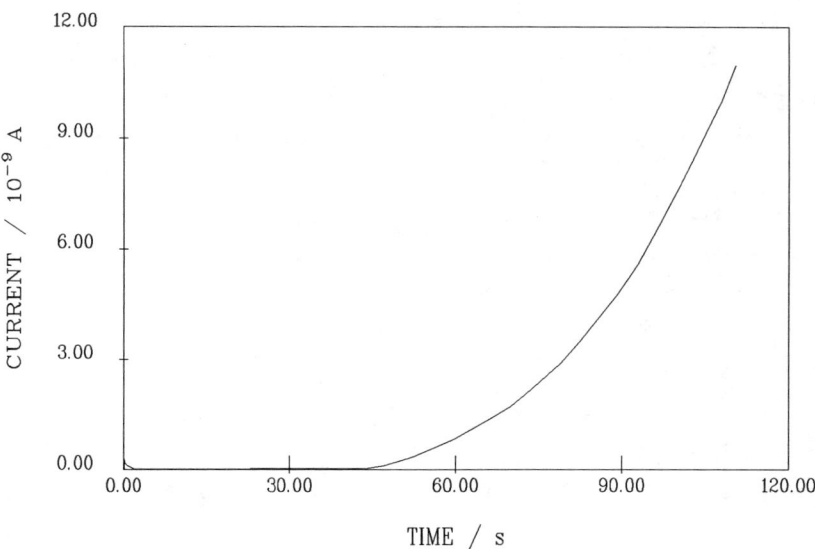

Figure 3. Current time transient for the deposition of α-PbO$_2$ onto an 8μm C-microelectrode.

kinetically controlled growth even at very high overpotentials. In consequence, it is possible to simplify the interpretation of the distributions of the arrival times of the first nucleus and to derive the kinetics of the initial stages of cluster formation (42). We see that in this example, as in the voltage gated transmembrane ion currents, the subsequent essentially deterministic processes allow the observation of the initial stochastic events so that it becomes possible to investigate electrochemical kinetics at the molecular level. We note finally that if the electrodes are sufficiently small, then it becomes possible to observe the higher moments of the reaction rates (the deterministic or stochastic mean being the first moment) e.g., see (29,26,43). The interpretation of the higher moments of the reaction rates is highly diagnostic of the mechanisms of the reactions; the measurement and interpretation of electrochemical "noise" therefore promises to be a further important area for the application of microelectrodes.

References

1. R. Lines and V.D. Parker, Acta Chem. Scand., B31 (1977) 369.
2. J.O. Howell and R.M. Wightman, J. Phys. Chem., 88 (1984) 3915.
3. A.M. Bond and T.F. Mann, in the press.
4. A.M. Bond, M. Fleischmann and J. Robinson, J. Electroanal. Chem., 168 (1984) 299.
5. J. Cassidy, S.B. Khoo, S. Pons and M. Fleischmann, J. Phys. Chem., 89 (1985) 3933.
6. A.M. Bond and P.A. Lay, J. Electroanal. Chem., 199 (1986) 285.
7. T. Dibble, S. Bandyopadhyay, J. Ghoroghchian, J.J. Smith, F. Sarfarazi, M. Fleischmann and S. Pons, J. Phys. Chem., 90 (1986) 5275.
8. M. Fleischmann, J. Ghoroghchian and S. Pons, J. Phys. Chem., 89 (1985) 5530.
9. M. Fleischmann, J. Ghoroghchian, D. Rolison and S. Pons, J. Phys. Chem., 90 (1986) 6392.
10. M.J. Peña, M. Fleischmann and N. Garrard, J. Electroanal. Chem., 220 (1987) 31.

11. A.M. Bond, M. Fleischmann and J. Robinson, J. Electroanal. Chem., 180 (1984) 257.
12. J. Ghoroghchian, F. Sarfarazi, T. Dibble, J. Cassidy, J.J. Smith, A. Russell, M. Fleischmann and S. Pons, Anal. Chem., 58 (1986) 2278.
13. A.M. Bond, M. Fleischmann and J. Robinson, J. Electroanal. Chem., 172 (1984) 11.
14. J. Kiwi and M. Gratzel, J. Am. Chem. Soc., 101 (1979) 7214.
15. D.S. Miller, A.J. Bard, G. McLendon and J. Ferguson, J. Am. Chem. Soc., 103 (1981) 5336.
16. W.J. Albery, P.N. Bartlett and A.J. McMahon in "Photogeneration of Hydrogen," A. Harriman and M.A. West, eds., Academic Press, London (1982).
17. W.J. Albery, P.N. Bartlett and A.J. McMahon, J. Electroanal. Chem., 182 (1985) 7.
18. R. de Levie in "Adv. Electrochem. Electrochem. Eng.," H. Gerisher and C. Tobias, eds., Wiley Interscience, 13 (1984) 1.
19. P. Mueller and D.O. Rudin, Nature, 217 (1967) 713.
20. L.G.M. Gordon and D.A. Haydon, Biochim. Biophys. Acta, 255 (1972) 1014.
21. M. Eisenberg, J.E. Hall and C.A. Mead, J. Membrane Biol., 14 (1973) 143.
22. M. Fleischmann, C. Gabrielli, M.J.G. Labram, A.I. McMullen and T.H. Wilmshurst, J. Membrane Biol., 55 (1980) 9.
23. M. Fleischmann, C. Gabrielli, M.T.G. Labram and T. Markvart, J. Electroanal. Chem., 214 (1986) 427.
24. M. Fleischmann, C. Gabrielli and M.T.G. Labram, J. Electroanal. Chem., 214 (1986) 441.
25. E. Neher, B. Sakmann and J.H. Steinbach, Tslueger's Arch., 375 (1978) 219.
26. M. Fleischmann, M. Labram, C. Gabrielli and A. Sattar, Surf. Science, 101 (1980) 583.
27. M. Fleischmann and H.R. Thirsk, Trans. Faraday Soc., 51 (1955) 71.
28. A. Bewick, M. Fleischmann and H.R. Thirsk, Trans. Faraday Soc., 58 (1962) 2200.
29. M. Fleischmann, K.S. Rajogopalan and H.R. Thirsk, Trans. Faraday Soc., 59 (1963) 741.
30. M. Fleischmann, J. Pattison and H.R. Thirsk, Trans. Faraday Soc., 61 (1965) 1256.
31. E. Budevski, W. Bostanoff, T. Witanoff, Z. Stoinoff, Z. Kotzewa and R. Kaischew, Electrochim. Acta, 11 (1966) 1697.
32. L. Kasper, Trans. Electrochem. Soc., 77 (1940) 353.
33. W.K. Burton, N. Cabrera and C.F. Frank, Phil. Trans. Roy. Soc., London, Ser A, 123 (1951) 299.
34. M. Fleischmann and H.R. Thirsk in "Adv. Electrochem. Electrochem. Eng.," P. Delahay and C.W. Tobias, eds., John Wiley, New York and London 3 (1963) 123.

35. M. Fleischmann and J.A. Harrison, Electrochim. Acta, 11 (1966) 749.
36. M. Fleischmann, J. Koryta and H.R. Thirsk, Trans. Faraday Soc., 63 (1967) 1261.
37. D. Pletcher and C. McCallum, J. Electroanal. Chem., 70 (1976) 277.
38. S. Toschev and I. Markov, Electrochim. Acta, 12 (1967) 281.
39. P. Bindra, M. Fleischmann, J.W. Oldfield and D. Singleton, Disc. Faraday Soc., 56 (1973) 180.
40. G. Gumawardena, G.J. Hills and B. Sharifker, J. Electroanal. Chem., 130 (1981) 99.
41. A.M. Bond, M. Fleischmann, S.B. Khoo, S. Pons and J. Robinson, Extended Abstracts 165th Meeting of the Electrochemical Society, May (1984), Electrochemical Society, Pennington, N.J.;
Indian Journal of Technology, 24 (1986) 492.
42. M. Fleischmann, L.J. Li and L. Peter, 15th International Power Sources Symposium, Brighton, September (1985); Power Sources Vol. 11 and to be published.
43. E. Budevski, M. Fleischmann, C. Gabrielli and M. Labram, Electrochim. Acta 28 (1983) 925.

Jurgen Heinze presented the following recent analysis to this conference. This algorithm is especially effective in the simulation of non-linear problems, as is evident from the rates of convergence and the stability of the method when applied to the disk geometry.

The use of ultramicroelectrodes ($R < 20\mu m$) in electrochemistry makes it possible to study very fast chemical reactions. The capacity of these electrodes is very low and thus it is possible to reach scan rates of about 10,000V/s and still obtain a reasonable signal to background ratio.

The theoretical treatment of the ultramicroelectrode is more complicated than for "normal" size electrodes ($R \approx 1mm$) as the mass transport to a finite planar area does not obey the model of semi-infinite linear diffusion.

The diffusion to a finite disk electrode should be resolved into two components, one of which is perpendicular to the disk and the other radial. The partial differential equations describing this diffusion process can be formulated without difficulty. Recently, some numerical methods for the solution of these equations have been given (1-4). Among the finite difference approximations the ADE, the hopscotch and the ADI methods are commonly applied. The ADE (1) method is an explicit one which is fast but unstable when $\lambda(=D\Delta t/\Delta R^2)$ exceeds the value of 0.25. The hopscotch method is also explicit but it is stable for all λ values which makes it perhaps the fastest of all methods (2). The ADI method, which will be described here, is implicit and also stable for all values of λ and it is more accurate than the hopscotch algorithm.

In the following we describe the digital simulation of the simple cyclic voltammetric experiment using an improved ADI alogrithm.

For a reversible redox couple

$$A + ne \rightleftarrows B$$

the following transport equations hold:

$$\frac{\partial C_A}{\partial t} = D_A \left[\frac{\partial^2 C_A}{\partial r^2} + \frac{1}{r} \frac{\partial C_A}{\partial r} + \frac{\partial^2 C_A}{\partial h^2} \right] \qquad [1]$$

$$\frac{\partial C_B}{\partial t} = D_B \left[\frac{\partial^2 C_B}{\partial r^2} + \frac{1}{r} \frac{\partial C_B}{\partial r} + \frac{\partial^2 C_B}{\partial h^2} \right] \qquad [2]$$

The C_A and C_B characterize the concentrations of the oxidized and reduced form of the redox system. The concentrations are given in dimensionless units referred to the bulk concentration of the electroactive species A. h and r represent the cylindrical space coordinates perpendicular and radial to the disk.

For cyclic voltammetry, the time dependent change of E(t) is determined by the scan rate $v = \Delta E/\Delta t$. Beginning with the initial potential E_i, the potential-time function is then:

$$E(t) = E_i - v \cdot t$$

At the switching potential E_λ the potential returns linearly to its starting value,

$$E(t) = E_\lambda + v \cdot t.$$

Assuming that a semi-infinite mass transport occurs in radial and normal direction to the electrode, the following boundary and initial conditions result:

$$D_A \frac{\partial C_A(r,0,t)}{\partial h} + D_B \frac{\partial C_B(r,0,t)}{\partial h} = 0, \quad (r \leq R, \, t \geq 0) \quad [3]$$

$$D_A \frac{\partial C_A(r,0,t)}{\partial h} = \frac{i}{nFA} = C_{A(h=0)} \cdot k_f - C_{B(h=0)} \cdot k_b \quad [4]$$

$$D_A \frac{\partial C_A(r,0,t)}{\partial h} = D_B \frac{\partial C_B(r,0,t)}{\partial h} = 0, \quad (r > R, \, t \geq 0) \quad [5]$$

$$\lim_{r \to \infty} C_A(r,h,t) = \lim_{h \to \infty} C_A(r,h,t) = 1, \quad (t \geq 0) \quad [6]$$

$$\lim_{r \to \infty} C_B(r,h,t) = \lim_{h \to \infty} C_B(r,h,t) = 0, \quad (t \geq 0) \quad [7]$$

In both these equations k_f and k_b symbolize the potential dependent rate constant for both directions of charge transfer; R is the electrode radius. The following absolute rate expression describes the variation of k_f and k_b with variation in the potential E(t):

$$k_f = k_s \cdot \exp\left[-\frac{\alpha nF}{RT}\left\{E(t) - E^0\right\}\right] \quad [8]$$

$$k_b = k_s \cdot \exp\left[(1-\alpha)\frac{nF}{RT}\left\{E(t) - E^0\right\}\right] \quad [9]$$

where k_s is the heterogeneous charge transfer rate constant at the standard potential E^0 and α the charge transfer coefficient.

Using the improved ADI-technique for the digital simulation, the following equations hold for the radial implicit half step:

$$c_{i,j,l+1/2} - c_{i,j,l} = \left[\underbrace{\left(T_{1r} - T_{2r}\right)}_{\text{implicit}} + \underbrace{\left(T_{1a} - T_{2a}\right)}_{\text{explicit}}\right] \quad [10]$$

$$T_{1r} = \beta_{i,j} \frac{F^V_{i,j+1}}{\Delta r_{j+1}} \left(c_{i,j+1,l+1/2} - c_{i,j,l+1/2}\right) \quad [11]$$

$$T_{2r} = \beta_{i,j} \frac{F^V_{i,j}}{\Delta r_j} \left(c_{i,j,l+1/2} - c_{i,j-1,l+1/2}\right) \quad [12]$$

$$T_{1a} = \beta_{i,j} \frac{F^H_j}{\Delta h_{i+1}} \left(c_{i+1,j,l} - c_{i,j,l}\right) \quad [13]$$

$$T_{2a} = \beta_{i,j} \frac{F^H_j}{\Delta h_i} \left(c_{i,j,l} - c_{i-1,j,l}\right) \quad [14]$$

$$\beta_{i,j} = \frac{D \Delta t}{2 V_{i,j}} \; ; \; I > i \geq 1; \; J > j \geq 2 \quad [15]$$

For the axial implicit half step, the following transformation must be done:

$l \rightarrow l+1$

T_{1r}, T_{2r} become explicit [16]

T_{1a}, T_{2a} become implicit

When an expanding grid is used, much computer time and space can be saved without loss of accuracy:

Expanding function: $f(x) = A\, e^{px^2}$ [17]

$$V_{i,j} = \left(h_{i+1} - h_i\right)\left(r_{j+1}^2 - r_j^2\right)\pi$$

$$F_{i,j}^v = \left(h_{i+1} - h_i\right) 2 r_j \pi$$

$$F_j^H = \left(r_{j+1}^2 - r_j^2\right)\pi$$

$$\Delta h_i = \frac{h_{i+1} - h_{i-1}}{2} \;;\; i \geq 2;\; \Delta h_1 = 1/2\Delta H$$

$$\Delta r_j = \frac{r_{j+1} - r_{j-1}}{2} ; j > 1$$

$$h_i = h_{i-1} + \Delta H\, e^{(i-1)^2 p};\; i \geq 3$$

$h_0 = 0$; $h_1 = 1/2\Delta H$; $h_2 = \Delta H$

$$r_j = r_{j-1} + \Delta R;\; e^{(j-1-n_R)^2 p};\; j > n_R$$

$r_j = r_{j+1} - \Delta R;\quad n_R \geq j > 0$

$r_{n_R+1} = n_R \Delta R = R$; R: electrode radius

$\Delta H, \Delta R$: standard space values

p: expansion parameter

The boundary conditions [3] to [7] take the following form with our algorithm: (5)

i=I : $T_{1a} = 0$ (axial outer boundary) [18]
j=J : $T_{1r} = 0$ (radial outer boundary)
j=1 : $T_{2r} = 0$ (center of electrode)
i=0 : surface: no radial diffusion

$T_{1r} = T_{2r} = 0$

j>n_R : isolated surface

$T_{2a} = 0$

1≤j≤n_R : electrode surface

radial implicit:

$$T_{2a}^m = \frac{4D^m \Delta t}{3\Delta H^2}\left[-c_{2,j,1}^m + 9c_{1,j,1}^m - 8c_{0,j,1}^m\right]$$

m=A,B

axial implicit:

$$T_{2a}^A = \frac{4\Delta t}{\Delta H}\left[-k_f c_{0,j,1+1}^A + k_b c_{0,j,1+1}^B\right]$$

$T_{2a}^B = -T_{2a}^A$

Since the concentrations of both forms of the redox couple determine the flux at the phase boundary they have to be simultaneously considered in the computation. Therefore, the number of difference equations is doubled in comparison to the simple model of diffusion controlled current.

A total of n_r annular elements of width Δr cover the electrode. Therefore, for determination of the experimentally relevant total flux all fractional fluxes in these elements must be summed. Near the edge of the electrode, the flux changes nonlinearly with the index j. Therefore the outer n_s rings of the active electrode surface must be more closely defined. The concentration values in the range from F^v_{i,n_r-n_s+1} to F^v_{i,n_r+1}, $0 \leq i \leq 2$, are interpolated using a Spline-function. The result is concentration values $C^m_{i,k,1}$, m=A,B, $0 \leq i \leq 2$, $1 \leq k \leq n_f$, which are closer in n_f.

The total flux is

$$\Phi_{tot} = \frac{i}{nFA} = \left[\sum_{j=1}^{n_r-n_s} f^A_{j,1} f^H_j + \sum_{k=1}^{n_f} f^A_{k,1} \left(r^2_{k+1} - r^2_k \right) \pi \right] / nFA \qquad [19]$$

where

$$f^A_{k,1} = \frac{D_A}{3\Delta H} \cdot \left(-C^A_{2,k,1} + 9C^A_{1,k,1} - 8C^A_{0,k,1} \right) \qquad [20]$$

and

$$r_k = r_{n_r-n_s+1} + \left[(k-1)(r_{n_r+1} - r_{n_r-n_s+1}) \right] / n_f \qquad [21]$$

Typical values for n_r are 10 to 100, for n_s, 9 and for n_f, 100.

The dependence of the dimensionless current function $\chi(at)$ on increments in the potential difference $E(t)-E^0$ can be calculated on the basis of the total flux. As each time increment Δt of the simulation is correlated to an equidistant change ΔE of the electrode potential, there is an unequivocal relationship between the flux and the actual electrode potential $E(t)$. In terms of the theory of Nicholson and Shain the simulated current function takes the following form (6):

$$\pi^{1/2}\chi(at) = \Phi_{tot}/C_A^b \left[D_A a\right]^{1/2} \quad [22]$$

with $a = \frac{nF}{RT} \cdot v$, and C_A^b the bulk concentration of the species A.

The influence of radial diffusion on the cyclic voltammetric current-potential curves can be easily estimated with the aid of a variable correction term which, analogous to the modified Cottrell equation for the disk electrode, is added to the ideal current function

$$\pi^{1/2}\chi(at) = \pi^{1/2}\chi(at)_{ideal} + \sigma p(at), \quad [23]$$

where the dimensionless parameter σ combines D, R and a which are characteristic for the radial transport process:

$$\sigma = \left[D/aR^2\right]^{1/2} \quad [24]$$

References

1. J.B. Flanagan and L. Marcoux, J. Phys. Chem., 77 (1973) 1051.
2. D. Shoup and A. Szabo, J. Electroanal. Chem., 160 (1984) 1.
3. J. Heinze, J. Electroanal. Chem., 124 (1981) 73.
4. B. Speiser and S. Pons, Can. J. Chem., 60 (1982) 1352; 2463.
5. J. Heinze, M. Störzbach and J. Mortensen, J. Electroanal. Chem., 165 (1984) 61.
6. R.S. Nicholson and I. Shain, Anal. Chem., 36 (1964) 706.

Charles Martin contributed the following on his research.

R. M. Penner and I have been working on a new procedure for constructing arrays of ultramicrodisk electrodes (1). This procedure is simple, quick, and requires only routine, inexpensive electrochemical instrumentation. Furthermore, this procedure can be used to prepare arrays with very small and uniform element radii. In principle, element radii as small as 50Å are possible using this approach; the arrays described in this paper had elements of 1000Å and 5000Å. These are the smallest element sizes for ultramicrodisk electrode arrays to be reported to date. Details of this procedure can be found in reference (1).

The procedure stems from our work with electronically conductive composite polymer membranes (2,3). This procedure is summarized schematically in Figure 1. A porous host membrane is immobilized onto the surface of a Pt disk electrode (Figure 1a). This membrane-modified electrode is then immersed in a solution of chloroplatinic acid and Pt is deposited by electrochemical reduction in the pores of the host membrane (1). Electrochemical deposition is continued until the Pt layer begins to overgrow the surface of the host membrane (Figure 1b). The surface of the Pt/Nuclepore composite membrane is then impregnated with polyethylene (PE) by immersion in molten PE solution (see Figure 1c and further below). Finally, the PE and excess Pt are removed by polishing, revealing the ultramicroelectrode (Figure 1d).

An important point is that the pores of the membrane host act as templates for the elements of the ultramicroelectrode. The geometries and dimensions of the pores define the geometries and dimensions of the ultramicroelectrode elements. The pore density defines the ultramicroelectrode element density.

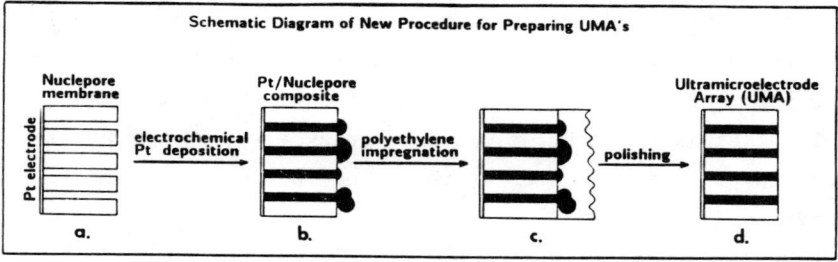

Figure 1. Schematic diagram of the procedure used to prepare ultramicrodisk electrode arrays.

Nuclepore polycarbonate membranes were used as the host materials (1). These membranes contain linear, cylindrical pores of nearly uniform pore diameter. This ensures that the elements of the ultramicroelectrode are all circular and of approximately the same diameter. Furthermore, because membranes with pore radii ranging from 6μm to 50Å are available commercially, ultramicroelectrodes with a broad range of element radii can, in principle, be prepared. In addition, these membranes have pure densities as high as 6×10^8 pores·cm^{-2}.

An electron micrograph of a typical ultramicroelectrode array is shown in Figure 2. Clearly Pt is deposited only within the pores of the host. Furthermore, the exposed Pt microdisks retain the size and near circular shape of the pores. Prior studies have shown that the Nuclepore host membrane is quite soluble in CH_2Cl_2 (2). Dissolving away the host membrane affords an opportunity for observing the Pt fibers which extend through the Nuclepore. Figure 3 shows, as expected, that these fibers assume the shape and dimensions of the host's pores. Note that the fibers in Figure 3 constitute an array of Pt microcylinders.

Savéant has suggested that the diffusion layer developed at an ultramicroelectrode during a voltammetric experiment can be divided into nonlinear (i.e., radial) and linear diffusion zones (4). If the direction perpendicular to the ultramicroelectrode surface is assigned the coordinate x, the nonlinear diffusion zone extends from x=0 (i.e., at the ultramicroelectrode surface) to a distance x=L, where L is approximately equivalent to the distance between the centers of the active sites on the ultramicroelectrode surface (4). The linear diffusion zone extends from x=L to some value x=Z, where Z>L. The exact value of Z depends on the time scale of the experiment (1,4).

The shape of the voltammogram will depend on the relative

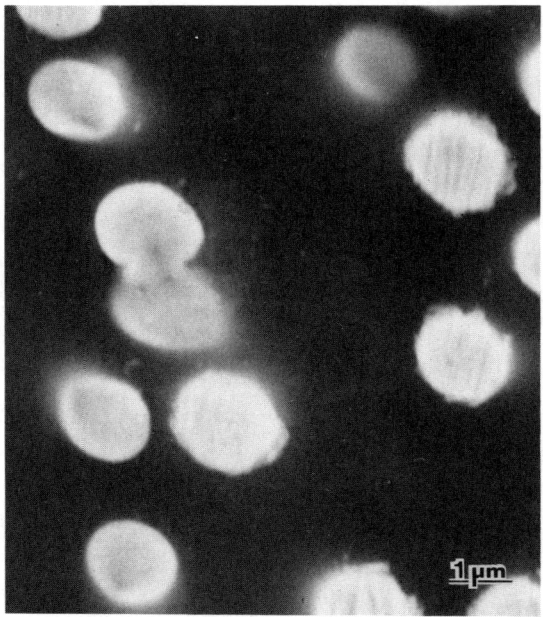

Figure 2. Electron micrograph of the surface of an ultramicroelectrode prepared from Nuclepore membrane with 1.0μm diameter pores.

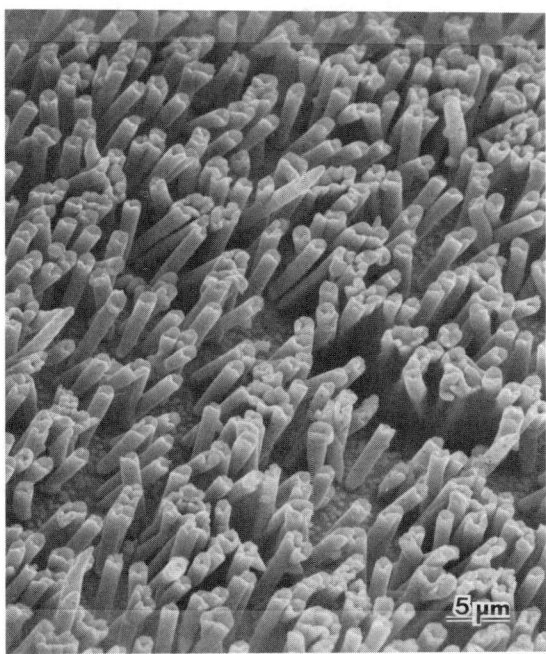

Figure 3. Electron micrograph of Pt fibrils obtained after polycarbonate host membrane is dissolved away from the ultramicroelectrode shown in Figure 2.

thicknesses of the nonlinear and linear diffusion zones. Consider the simplest limiting case (4). If the scan rate is very low, the linear diffusion layer will be much thicker than the nonlinear zone and diffusion will be dominated by transport in the thick linear zone. In this case, the voltammogram will look like a conventional semi-infinite linear diffusion voltammogram; indeed, currents observed will be identical to currents obtained for the same solution at a macro-sized electrode of equivalent geometric area (4,5). Because capacitive currents are proportional to the total geometric area, this limiting case will yield a signal to background advantage and is, therefore, of interest from an analytical point of view.

As scan rate is increased, the linear diffusion zone becomes thinner. When the scan rate is so high that the linear diffusion layer is much thinner than the nonlinear zone, a second limiting case is reached. In this case, current is dominated by radial diffusion in the nonlinear diffusion zone and the voltammogram assumes the shape of a polarogram (4). As is the case at an ultramicroelectrode, current is independent of scan rate when this limiting case is operative (1,4,5).

Cyclic voltammetric experiments show that the ultramicroelectrodes prepared conform to the theoretical considerations outlined above. Slow scan ($5mV \cdot s^{-1}$) voltammograms for Fe are shown in Figure 4. Note that the voltammograms obtained at the ultramicroelectrodes are identical to voltammograms obtained at a macro-sized electrode of the same geometric area. Thus, at this low scan rate, current is dominated by transport in the very thick linear diffusion zone and the linear diffusion to the total geometric area limiting case, discussed above, is operative. As far as we know, this is the first time that this analytically useful limiting case has been demonstrated

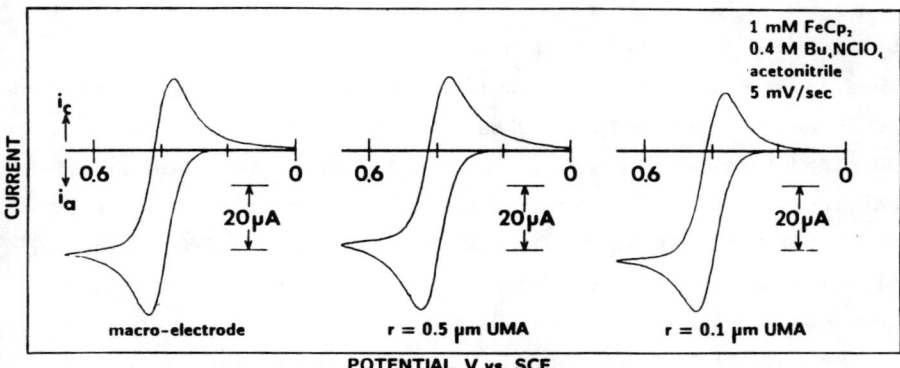

Figure 4. Cyclic voltammograms at 5mV/s in 1mM ferrocene, 0.4M Bu_4NClO_4, acetonitrile for typical r=0.1µm and 0.5µm ultramicroelectrodes for a macro-sized Pt electrode (A=0.5cm^2).

experimentally.

Figure 5 demonstrates the effect of scan rate on the shape of the voltammogram at the $0.5\mu m$ ultramicroelectrode. As predicted above, as scan rate increases, the voltammograms begin to look like polarograms. Note, however, that even at the highest scan rate investigated ($100V \cdot s^{-1}$), the wave still shows a cathodic peak. At higher scan rates, the voltammogram become distorted by slow electron transfer and uncompensated resistance effects. Thus, pure polarographic behavior (the second limiting case) could not be achieved with these arrays.

The experiments discussed above show that the electrochemical responses of the ultramicroelectrodes prepared here are in agreement with established theory (4,5). Quantitative evaluation of these ultramicroelectrodes using a theoretical analysis developed by Pons et al. (6) is in progress.

We have thus developed a simple and inexpensive method for preparing ultramicroelectrodes. This method can in principle produce ultramicroelectrodes with elements having diameters as small as 50Å. We are currently pursuing this possibility. In the most general sense, we describe in this paper a procedure for preparing uniformly sized and shaped metallic (or other material) microstructures. Such microstructures might be useful in a variety of chemical applications. For example, we have recently shown that electronically conductive polymer membranes having microfiberous morphologies can be prepared using the methods described in this paper (3). These microfiberous membranes support higher rates of ion-transport than conventional, amorphous electronically conductive polymer membranes (3). This has important implications for battery and other electronic applications of these polymers (9,16). It seems likely that unique catalyst structures could be prepared using the techniques discussed here (Figure 3).

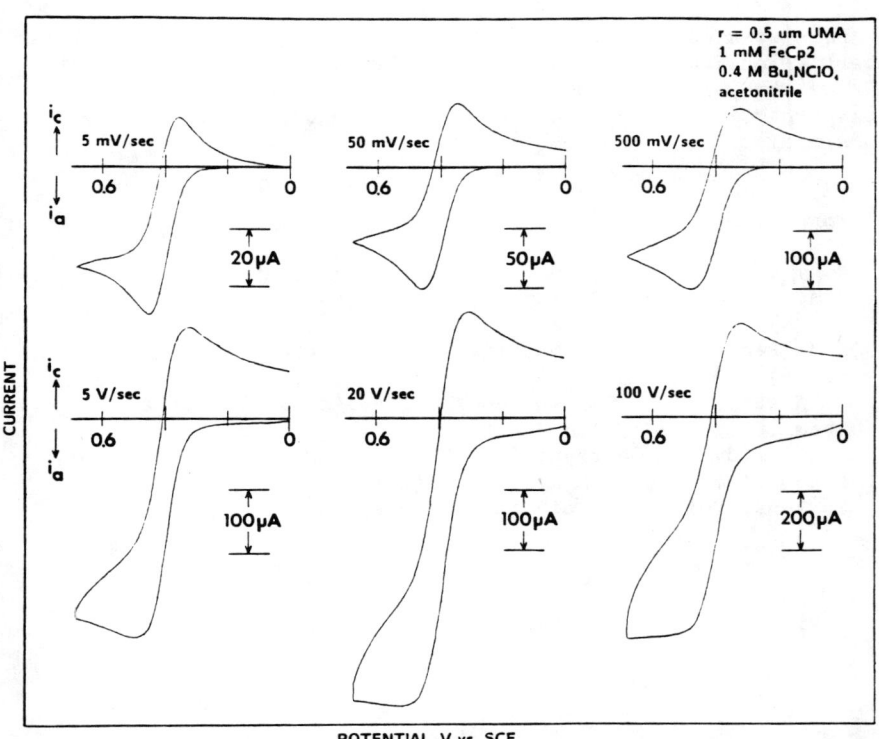

Figure 5. Cyclic voltammograms at various scan rates for a typical r=0.5μm ultramicroelectrode in 1mM ferrocene, 0.4M Bu$_4$NClO$_4$, MeCN.

Acknowledgements

This work was supported by the Office of Naval Research, the Robert Welch Foundation and the Dow Chemical Company. We would like to thank the Nuclepore Corporation for donating the Nuclepore membranes and Stan Pons for valuable advice and consultation.

References

1. R.M. Penner and C.R. Martin, Anal. Chem., submitted.
2. R.M. Penner and C.R. Martin, J. Electrochem. Soc., <u>133</u> (1986) 310.
3. R.M. Penner and C.R. Martin, J. Electrochem. Soc., <u>133</u> (1986) 2206.
4. C. Amatore, J.M. Saveant and D. Tessier, J. Electroanal. Chem., <u>147</u> (1983) 39.
5. J. Cassidy, J. Ghoroghchian, F. Sarfarazi, J.J. Smith and S. Pons, Electrochimica Acta., <u>31</u> (1986) 629.
6. S. Pons, University of Utah, in preparation.

Royce Murray presented the following overview of ultramicroelectrodes. His lecture was a review on the field of small structures in polymer electrodes, systems, chemical reaction kinetics, determination of thermodynamic parameters, and arrayed systems in electrochemistry.

It seems evident that the topic of ultramicroelectrodes comprises a collection of small electrode situations. I would divide these broadly into:

Small size

Some ultramicroelectrodes are electrodes of small dimension and varied shape (disk, hemisphere, band, ring, hemicylinder, appropriately partially illuminated semiconductor disk, etc.) mounted in or on some insulating support, and used under various conditions of voltammetric control. This Workshop has revealed a substantial effort both in fabrication of electrodes and in formulating the theoretical framework for mass transport to and various chemical and electrochemical kinetic complications at such electrodes. This fabrication and theoretical development is necessary and important, but it will be ultimately sterile unless there is commensurate application of these electrodes to new forms of chemical and electron transfer reactivity, of analysis, and of electrochemistry (not just verifying theory with model systems) that are simply not accessible with conventional electrodes and methods. This is one of the foci of my own research. There are important opportunities for new chemistry with ultramicroelectrodes, in low dielectric media, in media with ultralow ionic strength, and in viscous liquids and semi-solids, and regarding new levels of kinetic sensitivity and new types of chemistry coupled to heterogeneous charge transfer. It will require some chemical insights on the part of initial workers in the field as to what are the problems for which the microelectrode approach yields the crucial advantage; these are the most important chemical and electrochemical problems to pursue if the microelectrode strategy is to achieve widespread usage among a broader range of chemists.

Finely divided particulate suspensions

A second class of ultramicroelectrodes is finely divided particulate suspensions (or colloids) of electron transfer active or catalytically active materials (including semiconductor particles, zeolites, clays, metal microspheres, conducting oxides, etc.). Control or initiation of their electron transfer activity can be accomplished in three basic ways: photolytically by adsorption of light by the particle or by sensitizers in the homogeneous phase, electrochemically by electron transfer mediation with an electrode plus chargeable particles or redox species, and electrostatically by generation of a dipolar particle with a large electric field gradient. While the control/initiation step and the details of the particle's surfaces are made difficult to characterize precisely or manipulate (as compared to "ordinary" electrodes), one gains the important advantage of approaching the three-dimensionality of homogeneous solution reactions. This is already recognized and established as an important field of research. How cleanly connections can be made between it and that of "ordinary" electrodes, in terms of electrode reactions and associated chemistry, will be highly variable, but this is an important consideration.

Arrays

The third class of ultramicroelectrodes refers to arrays, or situations where there are multiple, small dimensioned electrodes of various shapes and interelectrode spacing. Both irregular and regular (patterned) arrays are known. Regular arrays are being approached using the highly developed microlithographic methods of the electronics industry, and with these one has the possibility of more than one working electrode (in fact tens or hundreds are conceivable). The arrays may or may not interact in terms of their mass transport fields; this matter is beginning to be addressed with theory and experiment. The potential diversity of the physical arrangement of arrays and the types of chemical and electrochemical studies to which arrays can be applied is considerable, and this complicates simple characterization of the subject. As one example, I anticipate a great future impact of array electrodes for analytical sensors. It is furthermore not illogical to consider a sensor with associated potentiostat, power supply and readout terminals on a single "chip" which is so inexpensive that it is used once or a limited number of times and discarded. That is, there is the potentiality of electrochemical methods becoming widely, not just occasionally, used for analysis. How about a personal CO sensor, or a sensor for the humidity over your drying-out evening roast?

Keith Oldham presented the following lecture on the role of mass transport and cell resistance in ultramicroelectrode behavior.

Microelectrodes have many disadvantages, as detailed in Figure 1. Despite these severe drawbacks, microelectrodes constitute an increasingly valued component of the electrochemist's repertoire because of the many unique advantages that are exhibited by electrodes of very small size.

The main advantages of microelectrodes are summarized in Figure 2 and are seen to center around two features that arise directly from their small size. One of these features is the large ratio of faradaic current to nonfaradaic current. The second beneficial feature of microelectrodes is their small resistance to area ratio. Both of these features arise directly from geometric considerations. Before addressing these in more detail, it is convenient to classify the types of microelectrode that are in use today.

Fabrication techniques often lead to electrodes of an "inlaid" configuration. By this is meant that the electrode material is imbedded in a surrounding insulator that forms a geometric continuation of the electrode surface. The shapes of such inlaid electrodes may be simple, as in the disk electrode, or complex, as with irregularly shaped inlays. A large class of inlaid microelectrodes, known variously as "band electrodes", "line electrodes", or "ring electrodes" have the general shape of a strip (with the ends joined in the case of rings) so that one dimension of the electrode far exceeds the other. Figure 3 illustrates some of the geometries involved. Because of mathematical difficulties, inlaid electrodes are often modelled as microhemispheres or microhemicylinders. As well, there are other electrodes, such as mercury microdroplets and microwires, for which mathematical treatment as hemispheres or cylinders is

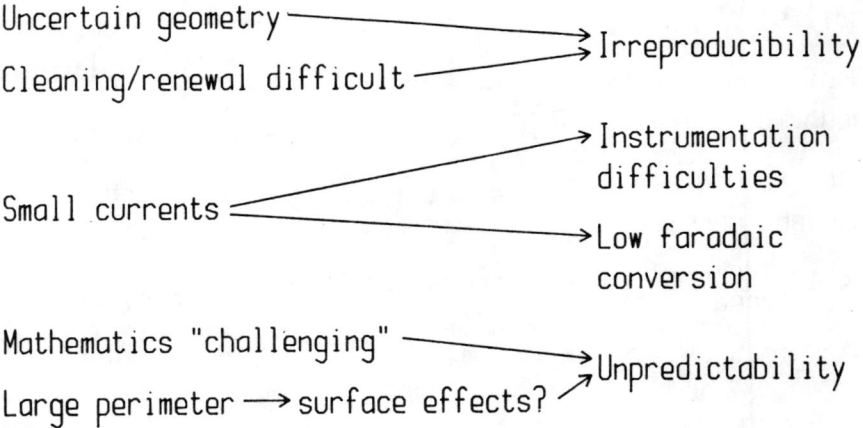

Figure 1. Microelectrode disadvantages.

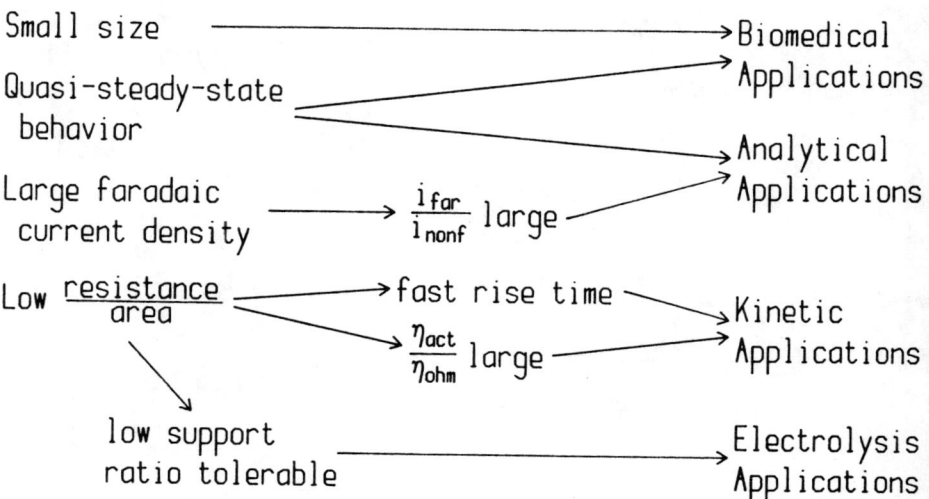

Figure 2. Microelectrode advantages.

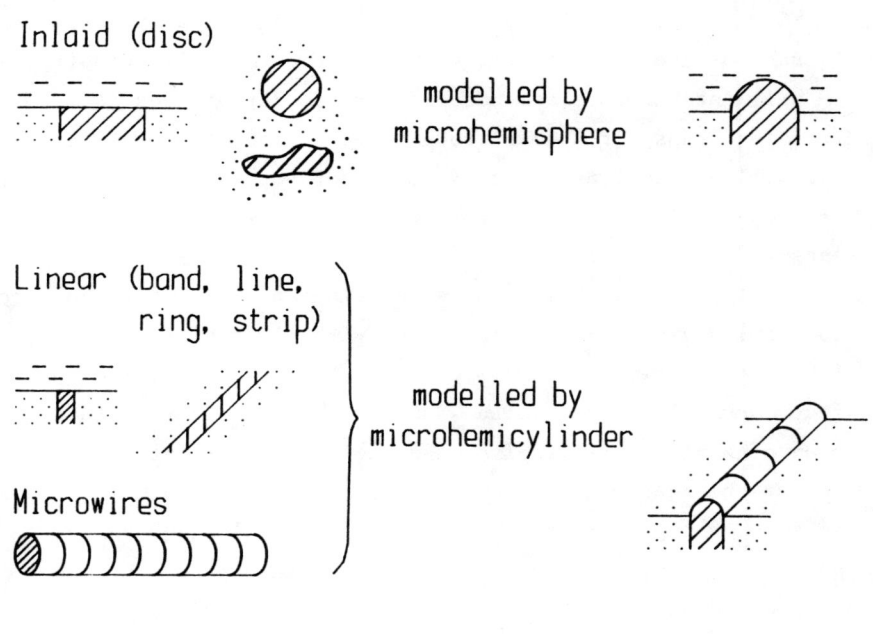

Figure 3. Types of microelectrodes.

more realistically appropriate. Array electrodes will not be discussed.

Though there have been many brilliant and exact treatments of transport to microelectrodes, these are generally complicated and will not be described here. Instead, let us look at diffusive transport to electrodes in a semiquantitative fashion that will expose the special features of very small electrodes. The concept of "electrode dimensionality" is useful in this respect.

An electrode is a surface separating an electronic conductor from an ionic conductor. In common with all finite surfaces, an electrode has two length dimensions. For example, a rectangular electrode has a length and a width, while an electrode in the shape of a ring has the circumference (or the diameter) of the ring as one length dimension and its width as the other. If we let x and y represent these two linear dimensions, the area A of the electrode will equal the product xy, multiplied by some numerical constant, whatever its shape. Where diffusion occurs to an electrode there is another, nongeometric, dimension, the diffusion length d equal to $\sqrt{Dt}$, where D is the diffusion coefficient of the electroactive species and t is the duration of the experiment.

The dimensionality of an electrode depends upon the relative magnitudes of the dimensions x, y and d. An electrode is two dimensional (2D) if both x and y exceed d: conventional macroelectrodes are in this category. A one dimensional (1D) electrode is characterized by having one of the geometric dimensions, say x, longer than d but the other, y, less than d. For example, a strip electrode of 1cm length and 1μm width would usually fall within this category for t = 10s. Electrodes have zero dimensionality (0D) when x and y are both smaller than d: such electrodes are essentially points on the length scale of diffusion. Figure 4 summarizes these relationships. Because time is involved in the definition of d, a microelectrode's dimensionality will generally evolve as the experiment proceeds. At sufficiently short times, all electrodes are two dimensional; the dimensionality may progress through unity and eventually reach zero.

As the penultimate equation in Figure 4 shows, the current i passed by an electrode following a potential step is proportional to a characteristic length L. The current density I is therefore proportional to L/xy. The table in Figure 5 shows that L and L/xy depend crucially on the electrode dimensionality. For two-dimensional electrodes both these quantities show proportionality to 1/d: the typical cottrellian dependence on the reciprocal square-root of time.

Electrodes of unity and zero dimensionality display radically different behavior, resulting from the convergent nature of diffusion to microelectrodes of these geometries. As the final entry in Figure 5 shows, the current density at an electrode of zero dimensionality is time-independent and inversely proportional to the square root of the electrode area: in other words, the current density may be increased indefinitely by making an 0D electrode smaller and smaller! An electrode of unity dimensionality behaves in a less simple fashion: the current does not reach a steady state in this case (though time-dependence is small) and the current density depends in the tabulated logarithmic way on the diffusion length d and the lesser of the two linear dimensions, y. As a 1D electrode is made narrower and narrower, its current density again increases without limit, though less dramatically than for the 0D case.

Nonfaradaic currents at microelectrodes are proportional to electrode area and the faradaic/nonfaradaic current ratio can

Electrode has two linear dimensions: x, y Area $\approx$ xy

Also diffusion length $d=\sqrt{Dt}$

2D Electrodes (macro) $x \gg d \ll y$

1D Electrodes (linear) $x \gg d \gg y$

0D Electrodes (point) $x \ll d \gg y$

Current following step $= i = nFc^{b}D\left(\begin{matrix}\text{characteristic}\\\text{length}\end{matrix}\right) \propto nFc^{b}DL$

Current density $= I = \frac{i}{A} \propto nFc^{b}D\left(\frac{L}{xy}\right)$

Figure 4. Electrode dimensionality.

	L	L/A
2D Electrodes (macro)	$\dfrac{xy}{d}$	$\dfrac{1}{d}$
1D Electrodes (linear)	$\dfrac{x}{\ln(d/y)}$	$\dfrac{1}{y\ln(d/y)}$
0D Electrodes (point)	$\sqrt{xy}$	$\dfrac{1}{\sqrt{xy}}$

Figure 5. Characteristic lengths.

therefore be made arbitrarily large, for 0D and 1D electrodes, by making the electrode dimensions small enough. This is the salient advantage of microelectrodes for analytical applications.

As revealed in Figure 2, other applications of microelectrodes benefit from the small ratio of resistance to area. Again this arises from the enhanced transport, this time the migration of charge-carrying ions, to electrodes of 0D and 1D type, resulting from their convergent geometries. The conductance G of cells containing an electrode of area xy is equal to the product of the specific conductivity K and a characteristic length Δ. This latter is generally different from L and, for the three electrode dimensionalities, adopts the values tabulated in Figure 6. The quantity ℓ that appears in this table is the distance from the microelectrode to the counter electrode or, in the case of a three-electrode configuration, to the luggin of the reference electrode. It is this length that replaces d in defining electrode dimensionality in the context of resistance effects.

The equations in Figure 6 show that ohmic polarization, the familiar "iR drop" of electrochemists, is proportional to the ratio L/Δ of the two characteristic lengths. Studies aimed at measuring the kinetics of electrode reactions are impeded by ohmic polarization which obscures the activation polarization that reflects finite kinetics. Hence the greater the ratio of ohmic to activation polarization, the worse is the situation for kinetic investigations. Activation polarization depends on current density, so that a large value of iR/I represents an impediment to kinetic studies. Figure 6 shows that this ratio is proportional to the electrode area divided by Δ and lists values of the A/Δ quotient for the three electrode dimensionalities. We see that for electrodes of zero dimensionality

$$\text{Conductance } G = K\begin{pmatrix}\text{characteristic} \\ \text{length}\end{pmatrix} = K\Delta$$

$$\text{Ohmic polarization} = iR = \frac{i}{K\Delta} = \frac{nFc^bD}{K}\frac{L}{\Delta}$$

$$\begin{matrix}\text{Index of demerit} \\ \text{for kinetics}\end{matrix} = \frac{\eta_{ohm}}{\eta_{act}} \propto \frac{iR}{I} = \frac{\Lambda}{K\Delta}$$

	Δ	Λ/Δ
2D	$\frac{xy}{\ell}$	ℓ
1D	$\frac{x}{\ln(\ell/y)}$	$y\ln(\ell/y)$
0D	$\sqrt{xy}$	$\sqrt{xy}$

Figure 6. Resistance and dimensionality.

ohmic effects may be made arbitrarily small by making the electrode area small enough, whereas for macroelectrodes electrode area is irrelevant. The 1D case again involves logarithms: such electrodes benefit from small size but less dramatically than do 0D electrodes. These are the arguments that lead to the belief that it is advantageous to use microelectrodes in investigations of electrode kinetics.

The small magnitudes of the iR effect with microelectrodes permits electrochemists to tolerate lower conductivity, K, values and has led to experiments in which little or no supporting electrolyte is added. To understand exactly what the results of such experiments imply is a very difficult task. Some ions will inevitably be present as impurities even if none are added deliberately. Moreover, these ions will have a nonuniform distribution, even before the experiment commences, because the microelectrode will scavenge the surrounding solution to populate its double layer. The thickness of the double layer, the so-called Debye length, is large for a solution with low ionic content and will often be comparable to the size of the microelectrode. Thus, once the experiment starts, diffusion layer and double layer will be inextricably mixed and each will evolve in time, influenced strongly by the other.

Figure 7 illustrates that the ionic content of a microelectrode cell will increase as the experiment progresses. Even if the electroactive species is neutral, the electrode reaction must generate ions that will return to the bulk of the solution under the combined effects of diffusion, migration and double-layer factors. The counter electrode may provide other ions which may reach the environment of the microelectrode under the influence of the large fields that such experiments engender.

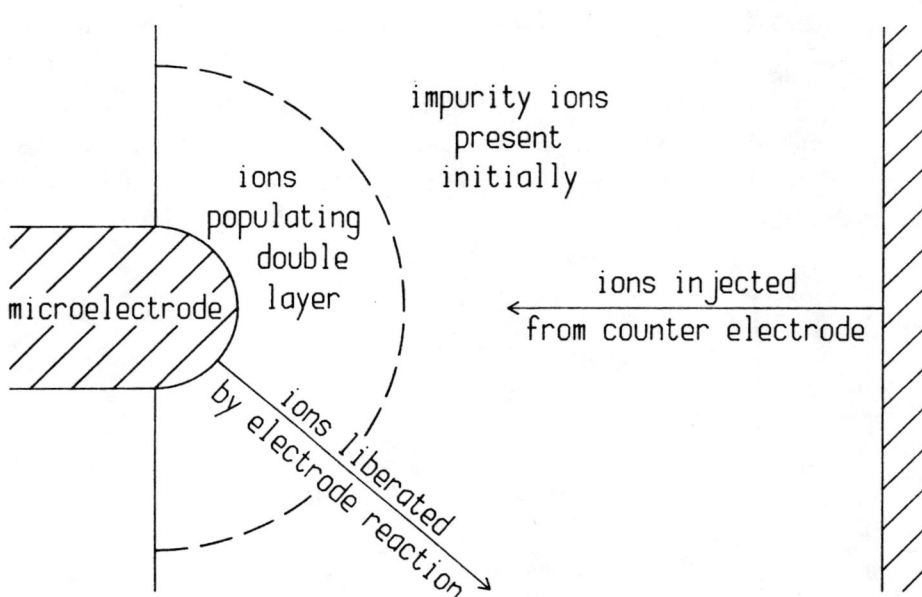

Figure 7. Ions in microelectrode cell.

A full mathematical treatment of microelectrode voltammetry under conditions of sparse supporting electrolyte represents a challenge that is unlikely to be overcome promptly. The cell as a whole, not just the microelectrode environment, will need to be modelled and calculations of the coupled fluxes must take into account diffusion, migration and lack of electroneutrality as a minimum. Other factors such as slow kinetics, solvent ionization, adsorption and ion pairing may well be important in certain cases, leading to further complications. The task is formidable.

Janet Osteryoung presented the following lecture on her work which is concerned with the use of very small electrodes for voltammetric analysis.

Chemical analysis has always relied on methods and procedures which can be used by nonexperts to yield reliable information about the nature and composition of chemical systems. The best foundation for such methods is an accurate and reasonably complete theoretical description. Lacking such a description it is either difficult or prohibitively expensive to identify the parameters on which the analytical signal depends and to specify the ranges of values of parameters for which the method should apply. In the context of very small voltammetric electrodes (i.e., those for which $r/\sqrt{Dt} < 1$, where r is a characteristic dimension and t is a characteristic time), the dependence of current on potential must be predictable for the geometry (inlaid disk, cylinder, etc) and the rate processes (diffusion, migration, charge transfer, etc) of interest. Thus the scope of analytical interest in theoretical problems involving small electrodes encompasses all of those generally thought necessary to fundamental characterization.

Problems of nonplanar diffusion are generally intractable mathematically. Their complexity makes it easy to make mistakes and difficult to estimate the effects of simplifying assumptions. In this context, it has been especially valuable to have several groups working on the same problem using different approaches. A good example of the approach to an accurate description of a simple case is given by Reference (1) and references therein. At the present state of understanding of very small electrodes, progress depends on continuing theoretical work with different approaches being employed by different workers. Because the results ultimately involve computation, it would be especially useful if there were more direct communication among the workers in this group via listings of computer programs or direct links to a computer network.

The second requirement for an analytical method is that the experimental system be well-defined and robust. The key component in a voltammetric experiment is the indicator electrode. For general voltammetric methods it is necessary for the geometry of the electrode to be well-defined and to be stable in use. It is also desirable for the geometry to conform to an ideal model so that it is practical to predict the response theoretically. It is worth pointing out parenthetically that most instrumental analytical procedures require an empirical calibration which is the weak point in the method. Coulometry, while formally instrumental, resembles titration and is automatically referred to as a primary standard. Voltammetry is an exceptional instrumental technique in that the response can generally be predicted from first principles. This great advantage should not be lost while trying to gain other advantages associated with small electrodes.

The slow pace of demonstrating and implementing the special properties of small electrodes has been set by practical difficulties in constructing electrodes. Many partially successful strategies have been employed (see (2) and references therein) but the central problem of converting a conductor into an electrode, that of sealing the conductor to an insulator, remains. This problem becomes more critical the greater the ratio of sealed edge to active area and applies to all strategies for fabrication which have been employed.

It is customary practice to polish the resulting electrodes mechanically when possible. This can coat the active surface

with a film of insulator (e.g., epoxy resin), distort the shape
of soft materials (e.g., gold), or damage the seal. This
practice often produces the desired result of a smooth, active
surface but is usually poorly defined and difficult to control
or specify. Because it is a critical part of preparing
electrodes, it requires equally critical examination.

Uncertainties in the electrode geometry as fabricated and
any change with use confound interpretation of voltammetric
results obtained with small electrodes. Microscopy, and in
particular optical microscopy, can and should be used routinely
to characterize geometry. The optical microscope has the great
advantage that it can be used to observe the electrode in the
presence of solvent. Classical optical interferometry can be
used to estimate a third dimension, while all the arts of
optical microscopy (e.g., using polarized or scattered light)
can be used to control preparation and treatment of electrodes.

Voltammetry implies measurement of current at various
potentials. The most popular way of conducting a voltammetric
experiment at a small electrode is to vary potential linearly
with time. This is a relatively easy experiment to carry out
but a difficult one to interpret (3). Pulse voltammetric
techniques are favored in cases where the limiting factor is the
capacitance of the electrode rather than the iR drop through the
cell. They will always give better results below some value of
concentration which depends on experimental details. Square
wave voltammetry (4) is especially useful for characterizing
electrochemical reactions at microelectrodes, because the shape
and position on the potential scale of the net current response
is largely independent of the geometry of the electrode or the
magnitude of the quantity $r/\sqrt{Dt}$. Thus, square wave voltammetry
can be used routinely to estimate reversibility and to measure
half-wave potentials using electrodes of irregular and uncertain

shape.

Although the inlaid disk is the most widely used geometry for small electrodes, it is the worst from the fundamental point of view in that the degree of nonplanar diffusion increases with increasing ratio of perimeter to area (for a two-dimensional figure) whereas the disk has the minimum value for this ratio. To achieve a substantial increase in current density at moderate times (i.e., $r/\sqrt{Dt} \ll 1$ at 100ms), would require sizes on the order of one micrometer and hence currents on the order of 100pA/m$\underline{M}$. Substantially the same performance can be achieved with cylindrical electrodes, but since the electrode area depends on length as well as radius, the effects of nonplanar diffusion can be achieved while independently adjusting the length to obtain the desired currents in the concentration range of interest. In addition, because the perimeter of the seal between cylinder and insulator occupies a minor fraction of the area subject to nonplanar diffusion, imperfections in the seal have little effect on the current. Recent advances in theory for cylindrical electrodes make it easier to validate experimental results for simple systems (5). Cylindrical geometries may play an important role in practical analytical applications.

References

1. K. Aoki and J. Osteryoung, J. Electroanal. Chem., <u>160</u> (1984) 335.
2. W. Thormann and A.M. Bond, Anal. Chem., <u>58</u> (1986) 000.
3. K. Aoki, K. Akimoto, K. Tokuda, H. Matsuda and Janet Osteryoung, J. Electroanal. Chem., <u>170</u> (1984) 99.
4. Janet Osteryoung and J.J. O'Dea, J. Electroanal. Chem., in press.
5. K. Aoki, K. Honda, K. Tokuda and H. Matsuda, J. Electroanal. Chem., <u>186</u> (1985) 79.

Stanley Pons lectured on the following work on applications of ultramicroelectrodes to electrochemistry in the gas phase and in highly resistive media.

New and interesting applications of ultramicroelectrodes have been the center of attention for many of us recently. High rates of mass transport to, and reduced ohmic and capacitative effects at these devices allow electrochemical measurements under conditions that eliminate the need for traditional electrochemical environments, such as conducting electrolyte solutions. Fleischmann, Bond and others have reported electrochemical experiments in liquid solutions containing little or no purposely added supporting electrolyte (1-8). We have found that the electrochemical potential window is generally extended in many experiments: acetonitrile can be used in simple redox reactions to about 4 volts vs a silver reference, and under special circumstances can be used to almost 6 volts. Under these conditions, we have been able to investigate the electrochemistry of species with very high vertical ionization potentials, including short chain alkanes, oxygen, nitrogen, and the rare gases (6). Electrochemical investigations in low temperature solution, solid eutectic mixtures, and glasses of nonaqueous solvents has also been reported (7): electrochemistry in these media permits studies of thermally stabilized reactive intermediates which may be generated subsequent to the electron transfer. Each of the above examples are possible because of the greatly reduced ohmic losses at ultramicroelectrodes.

Very small electrodes require only a small number of ion/dipoles to charge the electrical double layer. The required number may easily be supplied from the free carrier population present in solutions as impurities or from the autoionization of the solvent. Some of the aforementioned examples make use of this property. As a further novel example, we have recently used ultramicroelectrodes in the gas phase, and have demonstrated their use as an electrochemical detector in gas phase chromatography, as well as a sensitive electrochemical gas sensor. In these experiments, the charge source required for the double layer are ions present at the surface of the electronic insulator separating the two electrodes: an ultra-microelectrode and a large surface area reference electrode. The ultramicroelectrode itself is a thin ring or a small disk. Limiting currents measured are typically on the order of 10^{-10} to 10^{-16} amperes, in contrast to those measured at conventional macroelectrodes ($>10^{-9}$ amperes). Since the resistivity of the insulators used to mount the electrodes are on the order of 10^9 ohms, adequate carriers are present for double layer charging. We have demonstrated that the surface conductivity is also sufficiently high to allow measurement of Faradaic currents in the gas phase. These experiments must be distinguished from those that have been previously described where macroelectrodes in solid electrolytes have been used. In the present case, a true flux steady state is reached in short times. Ohmic potential losses are small, and the current response can be described in a manner similar to that in conventional electroanalysis. Thin gold ring ultramicroelectrodes (200-1000Å) were prepared by thermal reduction of reducing dilute solutions of metal organnometallics on a 5μm quartz fiber. A quartz or glass was used for the outer housing and reference electrode; the outer surface of the tube was coated with a thick layer of palladium by the same technique. The ultramicroelectrode was mounted in the palladium coated tube with epoxy resin. The two electrodes were thus separated by a thin layer of epoxy resin, and the wall thickness of the glass tube. Carbon and platinum disk ultramicroelectrodes (1-5μm diameter) were constructed in the same configuration by mounting fine wires in the glass tubes with epoxy resin. The finished ultramicroelectrode assembly was mounted in a Faraday cage to reduce the capacitatively coupled noise. The tip of the ultramicroelectrode was inserted inside the outlet port of a gas chromatograph so that the effluent of the chromatograph passed over the tip of the ultramicroelectrode. When chromatograms were recorded, and the peaks from the electrochemical detector and a commercial thermal conductivity detector were compared after normalizing to a common peak intensity, there were no observed differences in the dispersions of the two peaks demonstrating the good time response of the device. The typical background current in the system when polarized at +3.0V can be made to be on the order of 30 femtoamperes. Using this system, it has been possible to detect ppm level analytes in the gas phase. If the detection on equal injections is carried out as a function of potential, it is possible to construct polarograms of the analyte, which demonstrates that it is indeed the electron transfer reaction to the analyte that is being measured. Indeed it is possible to separate species with identical retention times by measuring the band at different potentials.

The mechanism of ion conduction across the insulating separator surface is most likely surface diffusion of protons between the two electrodes. It is clear that redox reactions which involve the elimination of a proton give rise to much larger responses than those that do not. In addition, the stability of organic cation radicals to proton loss correlate to the magnitude of the response as well. We have investigated the effects of coating the insulator with various materials and other methods of surface modification (etching, etc.) to increase conductivity, and thus sensitivity.

We demonstrated the use of these devices as sensitive detectors in other highly resistive media such as pure solvents in liquid chromatographic detectors some time ago. In pure acetonitrile, for instance, it was possible to detect picogram mL^{-1} quantities of several of the DNA free bases. The magnitude of the currents measured were dependent on the potential applied, demonstrating again that the detector is species sensitive.

Acknowledgement

We thank the Office of Naval Research for support of this work.

References

1. M. Fleischmann, J. Ghoroghchian, and S. Pons, J. Phys. Chem., 89 (1985) 5530. J. Ghoroghchian, M. Fleischmann, and S. Pons, in preparation. A.M. Bond, M. Fleischmann, and J. Robinson, J. Electroanal. Chem. 11 (1984) 172.
2. J. O. Howell and R. M. Wightman, Anal. Chem. 56 (1984) 524.
3. A. M. Bond, M. Fleischmann, and J. Robinson, J. Electroanal. Chem., 168 (1984) 299.
4. A. M. Bond, S. B. Khoo, S. Pons, and J. Robinson, Ind. J. Tech., 24 (1986) 492.
5. J. Cassidy, S. B. Khoo, S. Pons, and M. Fleischmann, J. Phys Chem., 89 (1985) 3933.
6. T. Dibble, S. Bandyopadhyay, J. Ghoroghchian, J. J. Smith, F. Sarfarazi, M. Fleischmann, and S. Pons, J. Phys. Chem., 90 (1986) 5275.
7. A. M. Bond, M. Fleischmann, and J. Robinson, J. Electroanal. Chem., 180 (1984) 257.
8. M. Fleischmann, S. Bandyopadhyay, and S. Pons, J. Phys. Chem., 89 (1985) 5537.

Hannah Reller contributed the following on her work.

A new type of electrode is proposed: a thin vertical line on the periphery of a rotating cylinder. This constitutes a microelectrode in one dimension, and is "infinitely large" in the other dimension. Gold microelectrodes at three widths were examined: L=20, 90 and 200μm. The reaction studied is the reduction of ferrocyanide. It was found that the limiting current density i_L is proportional to w^n, where n is 0.455, 0.482 and 0.511 for the 20, 90 and 200μm, respectively, which is less than the value of 0.7 measured at a rotating cylinder electrode. Plots of $\log(i_L)$ vs $\log(L)$ at different angular velocities gave straight lines, with a slight decrease of the slope with increasing speed of rotation. The values of the slopes are close to the value of -0.33 obtained for other configurations of microelectrodes under flow conditions. It was shown in this work that the limiting current density at strip microelectrodes on a rotating cylinder is significantly larger than that obtainable either on a RDE or on a RCE. For the thinnest microelectrode used in the present study (20μm) the enhancement, compared to the RDE is by a factor of 30.

Debra R. Rolison contributed submitted the following on dispersions of ultramicroelectrodes.

My interests in ultramicroelectrodes focus on three areas:

1. Use microelectrodes to simplify the measurement of thermodynamic parameters in electrochemical systems; in particular, measure ion-pair association constants in CH_3CN in the absence of deliberately added electrolyte. The two one-electron reductions of dinitroaromatic molecules in CH_3CN are very susceptible to the presence of trace amounts of alkali ion, and it would be useful to make the $E^{o'}$-$[M^+]$ measurements without the added ion-pairing complication of the tetraalkylammonium cation. Microelectrodes should make calculating such constants straightforward.

Preliminary experiments with 25μm and 10μm Pt disk electrodes, 1m$\underline{M}$ 1,4-dinitrobenzene, and freshly opened Burdick & Jackson CH_3CN (UV grade) showed a single peak upon cycling into the reduction waves. This indicates that a film has formed on the electrode surface and that the reversible two one-electron processes have been disrupted without adding any electrolyte. The coating can be removed at potentials positive of ferrocene oxidation. The water content of the CH_3CN appears sufficient to obscure alkali ion-pairing measurements and more rigorous treatment of the solvent is indicated.

2. Use microelectrodes in low dielectric media as sensors for trace electroactive components. The Navy would like to have small and *in situ* sensors to determine wear products in fuels and lubricants; many of the wear products should be electroactive. Microelectrodes fulfill all of these requirements as they are very small and they allow electrochemistry to be performed in nonpolar solvents in an *in situ* mode. These

investigations will be performed jointly with Professor Graham
Cheek (US Naval Academy) and scientists of the Surface Chemistry
Branch, NRL.

3. Use zeolites as a template to synthesize metal or metal
oxide microstructures ranging in size from single atoms to
clusters/particles (1-10nm) in the cages of the zeolite. These
microstructures can be explored as microelectrodes in low
dielectric media using dispersion conditions and feeder
electrodes. This ability to vary the size and nature of the
metal/metal oxide microstructure offers electrochemists a chance
to probe electrochemical reactivity of particles which, due to
their small size, do not exhibit bulk properties.

I am especially interested in using this supported-
microelectrode approach to investigate zeolites at
electrochemical interfaces. Zeolites have molecular size-and
shape-specificity and are one type of inorganic lattice that has
been placed on electrode surfaces to explore this effect on
electrochemical processes. A very practical way to explore the
interaction of molecular-size control with electrochemical
control is to place electrodes on the zeolite (i.e., metal or
metal-oxide microstructures) and then use the electrode-modified
zeolite particles as electrodes in bipolar dispersion
electrolyses.

Preliminary experiments demonstrate that supported
conductors (in this case, either Pt on alumina or Pt on Type Y
zeolite) will respond as microelectrodes when used as dispersed
bipolar electrodes to electrolyze high impedance H_2O or benzene-
H_2O without added supporting electrolyte. Zeolite-supported Pt
more effectively electrolyzed benzene-H_2O (i.e., more current
for less voltage) than did alumina-supported Pt at identical
metal loadings. This successful combination of electrode-

modified zeolites and dispersion electrolysis will allow a practical study of the intersection and possible synergism of electrocatalysis and zeolite catalysis.

Reference

1. D. R. Rolison, R. J. Nowak, J. Ghoroghchian, S. Pons, and M. Fleischmann, "Proceedings of the Third International Symposium on Molecular Electronic Devices", F.L. Carter and H. Wohltjen, eds., Elsevier, Amsterdam, 1987.

Parbury P. Schmidt submitted the following contribution on theoretical aspects of ultramicroelectrodes.

Introduction

Until recently, much of what we believed about phenomena existing in the microscopic electrical interface was obtained by inference, that is, by macroscopic thermodynamic and kinetic measurements (1). Over the past two decades there has been a rapidly growing interest in trying to find experimental methods to map directly the microscopic structure of the electrode/ solution interface. Spectroscopic techniques have been exploited to try to obtain additional detailed information about the electrode interface by using molecules sensitive to their environment as interfacial probes (2). An electrode that is fabricated small enough to sense only a single electron, proton or ion-transfer event has immediate appeal and many advantages. Such ultramicroelectrodes now exist and can act as another probe of the microscopic double layer (3).

This section will explore the possible connection between electrochemical experiments and theory. Computer simulation of interfacial events appears to be a useful link of theory and experiment in the case of the ultramicroelectrode. Ion and electron transfer to the electrode are of special interest, although such simulations have not yet been carried out. My purpose is to suggest the means by which such calculations may be realized. In addition, I shall discuss possible experiments in which the ultramicroelectrode probes systems which are otherwise inaccessible to normal electrochemical analysis, *viz.*, the electrochemistry of nonpolar systems. The ultramicroelectrode should be able to determine the electrocapillary response of an electrode in extremely dilute electrolyte. This permits a probe of the role of the solvent in the absence of the strong influence of the electrolyte.

The simulation of electron transfer to microelectrodes

Ultramicroelectrodes can sample individual complete electron-transfer events (4). If the surface of the electrode is small enough, and if the concentration of electroactive solute is appropriate, the measurement of individual transfers can be guaranteed. The reasoning is as follows. A normal, macroscopic electrode accepts or donates electrons to large numbers of donors or acceptors even over a very short span of time. In a statistical mechanical sense, the macroelectrode measures ensemble-averaged transfers which appear collectively as currents. An ultramicroelectrode seems to sample individual elements of the ensemble which account, on average, for the current. Ultramicrosampling techniques may provide a direct connection between experimental observations and statistical mechanical methods for generating elements of the ensemble of states. This possibility needs to be verified.

At the molecular level, we would like to know much more about an electron (or ion) transfer reaction. A fair question to ask, therefore, is how does one see more of the details of the mechanism of transfer from the macroscopically obtained activation energy and pre-exponential factor? By experimentation and theory, naturally. Double layer studies, such as examination of the capacitance and charging of the electrode, and studies of adsorption are essentially thermodynamic measurements. These studies frequently illuminate the path followed in part by the actual dynamic electron or ion transfer. (Direct kinetic measurements are numerous.) In the last decade there has been increased interest in wedding spectroscopy with electrochemical thermodynamic and kinetic measurements. Spectroelectrochemical investigation, although potentially a powerful tool, is not generally feasible with ultramicroelectrodes, although ultrathin wires can be made into an essentially transparent mesh.

An ultramicroelectrode is a natural device to use for kinetic measurements. As previously stated, there may be a connection between the single electron-transfer event and statistical mechanical ensemble theory. Thus, it may be possible to connect the theory of interfacial electron transfer more directly with the observable kinetics of reactions.

A discussion follows of the possible connections between the measurement of electron transfer events with the use of ultramicroelectrodes and specific forms of simulation which are beginning to be developed for these events. The discussion is based on transition state theory. The simulation, based upon the Monte Carlo (5,6) method, does not add anything to the specific development of transition state theory, but particular quantities such as activation energies and rate constants can be calculated. The computation of these rate parameters can assist in the comparison of theory with experiment. In addition, it may be possible to generate sequences of electron transfer events in essentially the same manner as measured by the ultramicroelectrode. To begin, I consider the use of Monte Carlo methods to evaluate activation energies and rate constants.

I

A considerable effort has been expended to determine transition state configurations for simple gas phase reactions using accurate potential energy functions (7). In many cases, the determination of the transition state by means of a saddle point calculation can be carried out using energies determined directly by quantum mechanical calculations (8-14). These calculations typically apply to molecular systems in the absence of any other interactions with surrounding molecules. Moreover, temperature does not usually enter directly into the mathematical analysis to determine these transition states.

One point to note is the fact that a saddle point determination for a single molecule in the absence of solvent or other molecules of any kind is a calculation at 0°K. In principle, nothing prohibits consideration of an environment, namely solvent (15-19). A saddle point calculation can be carried out for the expanded system, but the size of the complete system cannot be larger than the capacity of the computer to handle the calculation. A second point to note is the fact that these calculations can be carried out with use of atom-centered cartesian coordinates of the reactants and products. Although it is possible in some instances to sort out a reaction coordinate as a normal mode of the system, it is not generally necessary (or feasible) to do so. The steepest descents and saddle point routines used to optimize the energy of the system work with specific configurations of the atoms of the reactants and products (18,19). The description of the reaction in terms of a particular variation of the system potential energy as a function of an identified reaction coordinate does not directly enter the analysis.

Consider a sequence of configurations of solvent and solute. Within each of these configurations, a saddle point calculation is carried out. If the sequence is suitably chosen, various quantities, such as the energy, can be averaged over the sequence. In this manner quantities of interest for the reaction can be extracted, such as the activation energy.

In fact, a sequence of configurations can be used to generate the average values of various quantities. The method uses a Monte Carlo simulation of thermodynamic systems as originally proposed by Metropolis, et al. (5) in 1953 and as has

been used frequently since then (6). Consider a collection of
particles which is constrained to move within the boundaries of
a volume (or area, if the simulation is two-dimensional (5)) and
which is subject to attraction and repulsion. Images of the
primary cell can be constructed which minimize the effects of
arbitrary boundaries (as in the "minimum image" approach) or
other techniques can be used (Ewald sums, etc.) (6). Each
particle in the primary cell is moved an arbitrary, but random
amount. With each move, the energy of the system is determined
and compared to the energy of the previous configuration. If
the move results in a lower energy for the new configuration,
the configuration is kept. The energy (and any other property
of the system which can be calculated for the ensemble) is
included in an average.

If the generation of a new configuration yields an energy
which is higher than that of the previous configuration, a test
is made to determine whether it is replaced by the previous
configuration or kept. The test is a simple one. The Boltzmann
factor, $\exp[-(E_{new} - E_{old})/RT]$ is compared to a random number
whose value lies between zero and unity. The new configuration
is kept only if the Boltzmann factor is greater than the
randomly generated number. Quantities associated with the
retained configuration are included in the averages. In this
manner, one generates a Markoff chain of events with which to
compute average quantities.

The Metropolis method is a routine for energy minimization.
This makes it difficult to simulate the transition state, which
is a nonequilibrium state of the system. The following
paragraphs sketch one modification of the Metropolis sample
which, in principle, may allow one to calculate a rate constant
from a collection of individual configurations representing the
transition state (15-19).

In order to calculate the rate constant of a reaction, it is necessary to know the value of the activation energy and the pre-exponential frequency factor. The value of the activation energy can be determined by carrying out an optimization of the initial state of the reaction and then, by means of a force bias, drawing the system through the transition state to the final state. This approach, however, does not sample the fluctuations of the system in the transition state. This sampling is necessary in order to determine the frequency factor. As a result, a more difficult calculation must be carried out. The distribution must be weighted to force the Monte Carlo routine to sample configurations in the neighborhood of the transiton state. I now consider the nature of this weighting with a particular emphasis on ion transfer. Some developments concerning electron transfer will be mentioned; however, they are untested at this time.

Given an analysis which can generate an ensemble of initial and transition states, the energies of these states can be determined by direct averaging: (18,19)

$$<V> = (1/N)\Sigma_i v_i ; \qquad [1]$$

in which v_i is the energy associated with the ith congifuration. The activation free energy is therefore simply $<V^\dagger> - <V_0>$, where $<V^\dagger>$ is the average free energy of the transition state and $<V_0>$ is the average free energy of the initial state. As the energies of the reactive solute are determined in the presence of the solvent, the activation energy determined in this fashion reflects all appropriate rearrangements of the solute and reorientation and rearrangement of the solvent necessary to activate the reaction. This alone, however, does not immediately give any information about the frequency of passage over the barrier or tunnelling through the barrier.

The transition state for a single configuration as well as for the ensemble of configurations of the reacting system is globally unstable in a mechanical sense. That is, there exists one and only one negative eigenvalue of the diagonalization of the Hessian matrix of second order derivatives of the potential energy function for the system (20). This condition on the eigenvalues of the Hessian matrix is used analytically to locate the saddle point on the potential energy surface (8-14). The location of the saddle point determines the transition state. It is possible, in principle, to take each configuration of the initial state, as generated in the Monte Carlo simulation, and find the transition state by means of a saddle point optimization. In practice, such a calculation would be unfeasible; it would take too much computer time. The alternative is to constrain the system somehow to sample the region of the transition state. The sample of transition states can then be examined to determine the frequency factor for the passage over the barrier or tunnelling through the barrier. For each configuration of the transition state, the reaction will proceed either by passage over the barrier or a tunnel transition through the barrier. Thus, the average frequency factor should accurately reflect the appropriate balance of barrier overcrossings to tunnel transfers. The major problem remaining is the determination of whether the system passes over the barrier or tunnels through the barrier for an individual configuration.

One solution to this problem is particularly applicable to simple atom or ion transfers (18,19). A reaction is generally followed by tracking the migration of a single, identifiable atom, such as hydrogen or the proton. The motion of this migratory species characterizes the reaction, and it can be shown that either local mechanical stability or instability of

this species exists in the field of any instantaneous
configuration of the surrounding atoms and molecules. One then
calculates the collective effect of the surroundings on the
second derivatives of the potential at the location of the atom.
If the diagonalization of this 3x3 matrix yields an eigenvalue
which is negative, the motion of the migrating atomic species is
unstable along the principal axis associated with the negative
eigenvalue.

Assume that a saddle point configuration has been found. If
the migratory species can occupy a position which is locally
stable in the transition state, then one assumes that it reached
the top of the barrier by a path which is both electronically
and vibrationally adiabatic. The migrating species will then
pass over the top of the barrier with a frequency

$$\kappa = 2\epsilon/\hbar \; ; \qquad\qquad\qquad [2]$$

where ϵ is the eigenvalue associated with the principal axis
that has a major component in the direction of the reaction. If
no principal axis lies in a direction which takes the migrating
atom to a final state, then the saddle point for the individual
configuration is not a contributor to the reaction being
studied; it is a saddle point for another reaction, a side
reaction, or it is a spurious transient state of the system.

If a test of the stability of the migrating species
indicates that it is *locally unstable* in a mechanical sense, one
assumes that the particle must tunnel through the barrier.
Tunnelling might take place from the initial to the final state
of the system. With atom transfer, the actual distance of
migration is usually so great that this type of tunnelling will
not be observed. Instead, one believes that the migrating atom
is carried adiabatically by fluctuations of the system to a

position close to the top of the barrier where it is possible for the particle to tunnel.

The calculation of the tunnelling contribution to the rate can be carried out with the use of potential energy functions specific for the interactions involved (18,19). This analysis builds on an analysis which has been tested with the ammonia inversion problem (21).

In the past, most calculations of the tunnelling rate have been carried out with the use of model one-dimensional potential barriers of particular geometric types, some of which are square, triangular or parabolic barriers (22). The height and width or other properties of the barrier are difficult to relate directly to atomic and molecular properties of the system; this means that the calculations are somewhat arbitrary. The use of model potential energy functions for the interaction of pairs of atoms or for the many-body effects of bond bending, while also arbitrary, is generally held to be a more chemically intuitive and acceptable alternative (21). The analysis to determine the transfer frequency associated with a tunnel transition makes use of a set of basis functions which are expanded about the locations of the initial and final resting points for the migrating atom or ion (19). [A similar argument can be used to move the center-of-gravity of an electronic charge distribution if an electron transfer is involved.] We have shown in a calculation of the inversion frequency of ammonia that it is possible to use the individual atomic pair potentials to describe the interaction over the path taken by the migrating ion or atom as it tunnels (21). By means of a variational calculation, it is possible to determine the splittings of the spectral levels, and for the kinetic analysis, the transition probability for the tunnel transfer.

Each configuration of the transition state generated in a

Monte Carlo routine allows a transfer by means of an over-the-barrier, vibrationally adiabatic transfer or by means of a tunnel transfer through the barrier (15-19). Let κ_a be the frequency which is associated with the over-the-barrier transfer, and let κ_d be the transfer rate for the tunnelling. Averaging over N configurations generated in the Monte Carlo cycle, the frequency factor for the reaction is (19)

$$\kappa = (1/N)\Sigma_i \kappa_i \; ; \qquad [3]$$

where κ_i is either κ_a or κ_d for the adiabatic or tunnel cases. In this manner, one obtains the activation energy and a pre-exponential transfer frequency which accurately reflects the balance between tunnelling and over-the-barrier transfers. Using the simple Arrhenius form of the rate constant--valid for a restricted range of temperatures--one can predict values for the rate constant:

$$k = \kappa \exp(-V_a/k_B T) \; ; \qquad [4]$$

where V_a is the activation energy difference $\langle V^\dagger \rangle - \langle V_0 \rangle$.

We have carried out initial calculations of this type for model systems. The initial calculation involved the simple inversion of a bent ABA molecule in a solvent of simple spheres (23). Although such a calculation is not able yet to be tested directly against experiment, by examining these simple systems first, the algorithms developed can be used to investigate simple, practical systems.

II

Atom transfer offers a simpler computer simulation of the type described above than does electron transfer. Because of the extreme delocalization of the electron, it is difficult to manipulate a system of electronic charges into a transition state configuration. Preliminary investigations of this problem with the use of the proton--which can exhibit a considerable amount of delocalization--suggest that it is possible to move the center of mass or charge as a Gaussian wavepacket to an optimum position in a saddle point (19). The proton has a certain "hardness" which the electron does not have, so, it is possible to see optimization to a saddle point for the proton in which a hard core is manifestly unstable along a particular coordinate but the delocalized portion of the system (in an analytical sense) is still stable. A configuration can be found which corresponds to a true saddle point even for the delocalized mass distribution. In that case, tests similar to those described above for the classical masses of ions and atoms can be applied to the proton and electron. The center-of-gravity of the mass or charge distribution is moved instead of the center of a definite hard particle.

III

Now, we compare a possible computer simulation and an actual form of current measurement at an ultramicroelectrode. First, the experiment.

Fleischmann and Pons (21) have used an ultramicroelectrode to register individual electron transfers. This particular experiment is designed to use a diode to emit light each time an electron enters the ultramicroelectrode. The sequence of pulses of light emitted from the diode are received by a photodiode tube, amplified and counted. By means of statistical arguments, which Fleischmann and Pons elaborate elsewhere in this volume, it is possible to extract a current.

The use of the Monte Carlo simulation and an applied force bias allows the system to evolve a sequence of transition states. The calculation will be costly in time, but on a small and simple enough system one ought to be able to duplicate experimental conditions. A random sequence of initial states is generated. From each initial state, the force bias is applied and the system drawn to the transition state. Time can be measured in terms of the number of steps from any particular initial state to the transition state. Once in the transition state, the electron can be assumed to pass to the final state with the act recorded as the contribution to the current. Tests and restrictions can be built in to limit the number of successful transfers from the transition to the final state.

Alternately, imagine a molecular dynamics calculation which operates in the same manner (25). Here time is a real variable in the analysis and enters automatically, but the difficulty lies in that the equations of motion must be solved. This can be more complicated for quantum mechanical systems; in the Monte Carlo calculation, the frame of reference used is that of equilibrium statistical mechanics. The point in favor of the Monte Carlo approach is the fact that analyses we are developing for the study of the reactivity of atom and ion transfers (15-19) can be extended to the electron transfer, as indicated above. A further extention to simulate this kind of experiment is straightforward, in principle; in practice, the extention may not be as simple. Nevertheless, it appears to be a reasonable calculation to attempt and offers the more direct contact with experiment.

IV

A final topic concerning the connection between interfacial theory and experiment has to do with the Monte Carlo determination of interfacial capacitances. For several years now, Valleau and his colleagues at Toronto (26,27) have pioneered the computer simulation of double layer phenomena in part to test the validity of various theories of the double layer capacitance. The Gouy-Chapman theory, (28,29) for example, is a precursor to the Debye-Hückel theory (30) in which the Poisson-Boltzmann equation is used to determine the distribution of potential and charge in the interface. In view of the extreme simplifications needed to establish this equation, the question of its range of validity cannot be avoided.

The most directly accessible electrochemical quantity is the double layer capacitance. A virtue of the Monte Carlo simulation is its usefulness in yielding the double layer potential and charge in a rigorous manner for a given, well defined model. Thus, it is possible to explore the effect even small changes in the system Hamiltonian operator (i.e., the total energy) can have on a measurable quantity, the capacitance. A strength of the microelectrode is surely its closeness, once again, in approaching the ideal conditions which are set in the Monte Carlo model. It may be possible, therefore, to carry out capacitance measurements in solutions of very dilute electrolyte. The theory and simulation of these experimental examples ought to be possible.

It seems to me that at the moment there are two exciting areas of electrochemistry which are growing rapidly and which offer much hope for testing and prodding the further development of theory. The first is the spectroscopy of the interface. The second is the microelectrode. Much of the discussion at this meeting has concerned the development of transport theory appropriate to these devices. I have attempted to bring attention to the possibility that the microelectrode may be a direct probe of the electron transfer event. The simulation of this event is possible. The microelectrode may forge a closer link between theory and experiment of the electron transfer step.

Acknowledgement

This work is supported in part by a contract with the Office of Naval Research, Arlington, Virginia.

References

1. A. J. Bard and L. R. Faulkner, Electrochemical Methods, Fundamentals and Applications, (John Wiley & Sons., New York, 1980).
2. K. Ashley and S. Pons, Trends in Anal. Chem., 5 (1986) 263.
3. See Bibliography of M. Fleischmann herein.
4. M. Fleischmann and S. Pons, in preparation.
5. N. Metropolis, A. W. Rosenbluth, M. N. Rosenbluth, A. H. Teller, and E. Teller, J. Chem. Phys., 21 (1953) 1087.
6. J. P. Valleau and S. G. Whittington, in Statistical Mechanics, Part A: Equilibrium Techinques (Modern Theoretical Chemistry, Vol. 5), B. Berne, ed. (Plenum Press, New York, 1977).
7. P. Pechukas, in Dynamics of Molecular Collisions, Part B (Modern Theoretical Chemistry, Vol. 2), W. Miller, ed. (Plenum Press, New York, 1977).
8. J. W. McIver and A. Komornicki, J. Am. Chem. Soc., 94 (1972) 2625.
9. A. Banerjee, N. Adams, J. Simons, and R. Shepard, J. Phys. Chem., 89 (1985) 52.
10. J. Simons, P. Jørgensen, H. Taylor and J. Ozment, J. Phys. Chem., 87 (1983) 2745.
11. C. J. Cerjan and W. H. Miller, J. Chem. Phys., 75 (1981) 2800.
12. S. Bell, J. S. Crighton and R. Fletcher, Chem. Phys. Letters, 82 (1981) 122.
13. S. Bell and J. S. Crighton, J. Chem. Phys., 80 (1984) 2464.
14. D. Poppinger, Chem. Phys. Letters, 35 (1977) 550.
15. P. P. Schmidt, J. Chem. Soc. Faraday 2, 80 (1984) 157.
16. P. P. Schmidt, J. Chem. Soc. Faraday 2, 80 (1984) 181.
17. P. P. Schmidt, J. Chem. Soc. Faraday 2, 81 (1985) 341.

18. P. P. Schmidt, J. Chem. Soc. Faraday 2, $\underline{82}$ (1986) 1399.
19. P. P. Schmidt, Proc. Indian Acad. Sci., Chem. Sci., $\underline{97}$ (1986) 233.
20. J. N. Murrell and K. J. Laidler, Trans. Faraday Soc., $\underline{64}$ (1968) 371.
21. P. P. Schmidt and S. S. Chang. J. Phys. Chem., $\underline{90}$ (1986) 4945.
22. R. P. Bell, The Tunnel Effect in Chemistry (Chapman & Hall, London, 1980)
23. P. P. Schmidt and L. Blum, J. Phys. Chem., submitted.
24. M. Fleischmann and S. Pons, this meeting.
25. J. J. Erpenbeck and W. W. Wood, in Statistical Mechanics, Part B: Time Dependent Processes (Modern Theoretical Chemistry, vol 6), B. Berne, ed., (Plenum Press, Now York, 1977); J. Kushick and B. Berne, ibid.
26. S. L. Carnie and G. M. Torrie, Advances in Chemical Physics, Vol. LVI (John Wiley & Sons, New York, 1984) pp.141-253.
27. G. M. Torrie and J. P. Valleau, J. Chem. Phys., $\underline{73}$ (1980) 5807.
28. G. M. Torrie and J. P. Valleau, J. Chem. Phys., $\underline{86}$ (1982) 3251.
29. G. Gouy, J. Phys (Paris), $\underline{9}$ (1910) 457.
30. D. L. Chapman, Phil. Mag., $\underline{25}$, (1913) 475.
31. P. Debye and E. Hückel, Z. Physik, $\underline{24}$ (1923) 185.

Benjamin Sharifker contributed the following analysis and criticism for the problem of overlapping concentration distributions at arrays of disk ultramicroelectrodes.

Consider first *one* planar microdisk on an insulating surface:

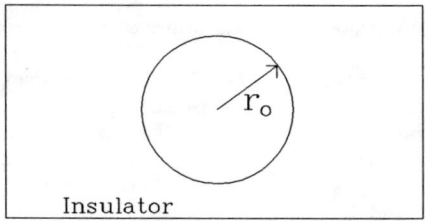

Insulator

Because of the hemispherical symmetry, diffusion towards the microdisk may be calculated from

$$\pi r_0^2 J_s(t) = -\pi r_0^2 D \left[\partial c(r,t)/\partial r \right]_{r=r_0} \quad [1]$$

where $\pi r_0^2 J_s$ is the total amount of material (i.e., area x flux) diffusing to the surface of the disk. For a disk of small dimensions, the inequality $(\pi D t)^{1/2} \gg r_0$ will hold for sufficiently long times. Equation [1] results in

$$\left[\partial \bar{c}/\partial r \right]_{r=r_0} = c/r_0 p + C/(Dp)^{1/2} \quad [2]$$

in the Laplace plane, and hence

$$\pi r_0^2 J_s(t) = -\pi r_0 Dc - \pi r_0^2 Dc/(\pi D t)^{1/2} \quad [3]$$

so that the amount of material $\pi r_0 J_s$ diffusing to the surface of an *isolated* microelectrode reaches a steady state. The steady state quantity as given by [3] at long times is in agreement with that given by Oldham, i.e., $\pi r_0^2 J_{s.s.} = Dc(P/2)$, where P is the perimeter of the inlaid microelectrode which, for a disk, is $2\pi r_0$.

In considering diffusion to an ensemble of microelectrodes the overlap of

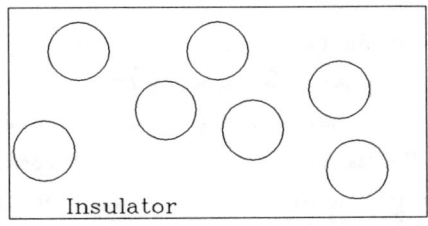

hemispherical diffusion zones around individual microelectrodes must be taken into account. At short distances from each microelectrode, diffusion to them will retain spherical symmetry but, as the distance becomes comparable with the intermicroelectrode separation, interaction between hemispherical diffusional fields occurs. At long times the only source of diffusing species is the bulk of the solution, i.e., the direction normal to the plane. A spherical-to-planar transition then occurs:

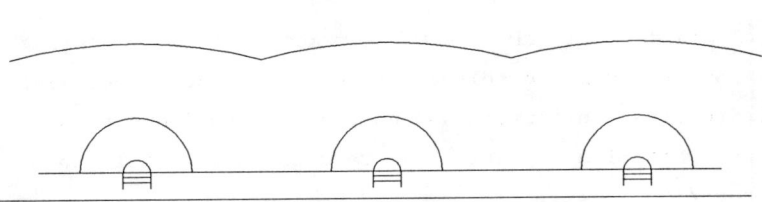

The treatment of the overlap of diffusional fields is complicated by their distribution in two dimensions (the plane of the ensemble) coupled with their growth in three dimensions, extending to the bulk of the solution. A considerable simplification is achieved by transforming the three-dimensional growth of the diffusional fields to an equivalent two-dimensional problem.

In spherical diffusion to a microelectrode, concentration gradients in directions other than that normal to the plane are not necessarily zero, and a steady state flux is eventually attained. In linear diffusion to a planar electrode, on the other hand, all gradients in directions other than that normal to the plane are zero, and thus no steady state flux in a stagnant solution occurs. If the segment of plane is allowed to grow, however, linear diffusion sustains the constant radial flux to the microelectrode surface. The radial-to-planar transition can be described then by considering the overlap of these equivalent segments of plane.

The equivalent area of a plane surface, πr_d^2, which results in the same amount of material diffusing through its surface as the amount that diffuses to a microelectrode through radial diffusion can be defined through

$$\pi r_d^2 J_p(t) = \pi r_d^2 D \left[\partial c(z,t)/\partial z \right]_{z=0} \qquad [4]$$

where $J_p(t)$ would be the flux at the surface of this equivalent diffusion zone and z is the direction normal to the plane. For semi-infinite linear diffusion and complete diffusion control in a stagnant solution, equation [4] results in the well known Cottrell equation:

$$\pi r_d^2 J_p(t) = -\pi r_d^2 Dc/(\pi Dt)^{1/2} \qquad [5]$$

The radius of the equivalent diffusion zone, r_d, is now obtained by equating the amount of material diffusing radially to the microelectrode (equation [3]), to the amount of material diffusing linearly to the equivalent diffusion zone, equation [5], i.e.,:

$$r_d^2 = \frac{(\pi Dt)^{1/2}}{\pi} = \left[\pi r_o + \frac{\pi r_0^2}{(\pi Dt)^{1/2}}\right] \qquad [6]$$

The area of an equivalent diffusion zone is then:

$$S_1 = \pi r_d^2 = \pi r_o (\pi Dt)^{1/2} + \pi r_0^2 \qquad [7]$$

The spherical-to-planar transition in diffusion to ensembles of microelectrodes can now be treated as the overlap of equivalent diffusion zones.

These types of arrays will be considered: square lattice, hexagonal lattice and random array.

(a) Square lattice

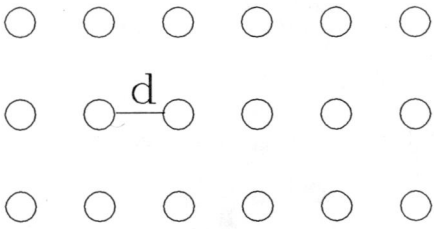

We may distinguish three regimes in the development of

diffusion zones:

(i) $r_d < d/2$, no overlap of diffusion zones.

The coverage of the array by diffusion zones is given by

$$S = \frac{\pi r_0 (\pi D t)^{1/2} + \pi r_0^2}{d^2} \qquad [8]$$

The number density of microelectrodes, n, is related to the intermicroelectrode distance, d, through

$$n = 1/d^2 \qquad [9]$$

Substituting [9] in [8]:

$$S = \pi \left[n r_0 (\pi D t)^{1/2} + n r_0^2 \right] \qquad [10]$$

We define now a dimensionless time u as

$$u = n r_0 (\pi D t)^{1/2} \qquad [11]$$

and note that the ratio of the area of the diffusion zones to that of the array is, when $r_d = d/2$,

$$S = \pi \left[u + n r_0^2 \right] = \frac{\pi}{4} \qquad [12]$$

Thus, the regime of no overlap of diffusion zones holds for times such that

$$u < \frac{1}{4} - n r_0^2 \qquad [13]$$

During the interval $0 < u < 1/4 - nr_0^2$, the current density to the array is given by the linear diffusion current to an electrode of fractional area S, i.e.,

$$I = \frac{FDc}{(\pi Dt)^{1/2}} \pi \left[nr_0 (\pi Dt)^{1/2} + nr_0^2 \right]$$

$$I = FDc \left[\pi\, nr_0 + \pi\, nr_0^2/(\pi Dt)^{1/2} \right] \qquad [14]$$

which can be expressed in units normalized to the current to independent microelectrodes as:

$$\frac{I}{\pi\, nr_0\ FDc} = 1 + \frac{nr_0^2}{u} \qquad [15.1]$$

(ii) $d/2 \le r_d \le d/(2)^{1/2}$, overlap of diffusion zones with microelectrode array partially covered by diffusion zones.

During the interval $1/4 - nr_0^2 \le u \le 1/2 - nr_0^2$, overlap of diffusion zones must be accounted for in the calculation of the current. Omitting the details of the calculation of the true coverage diffusion zones, the current is given, in normalized units, by:

$$\frac{I}{\pi nr_0\ FDc} = \frac{1}{\pi u} \left\{ \left[4\left(u+nr_0^2\right) - 1 \right]^{1/2} + 4\left(u+nr_0^2\right) \right\}$$

$$\left[\frac{\pi}{4} - \arctan\left[4\left(u+nr_0^2\right)-1\right]^{1/2}\right]\right\} \quad [15.2]$$

(iii) $r_d > d/(2)^{1/2}$, microelectrode array totally covered by diffusion zones.

For times such that $u > 1/2 - nr_0^2$, the electrode is totally covered by diffusion zones and the current to the array is given by

$$\frac{I}{\pi\, nr_0\, FDc} = \frac{1}{\pi u} \quad [15.3]$$

(b) Hexagonal lattice

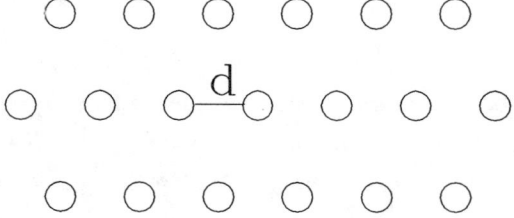

The number density of microelectrodes is in this case related to the intermicroelectrode distance by

$$n = \frac{2}{(3)^{1/2} d^2} \quad [16]$$

As with a square lattice, we identify three regimes of overlap:

(i) $r_d < d/2$, no overlap of diffusion zones.

In an hexagonal lattice this occurs for times such that $u < 1/2\,(3)^{1/2} - nr_0^2$. The current density is given by

$$\frac{I}{\pi\,nr_0\,FDc} = 1 + \frac{nr_0^2}{u} \qquad [17.1]$$

(ii) $d/2 \le r_d \le d/(3)^{1/2}$, overlap of diffusion zones, partial coverage of microelectrode array.

The interval of this regime is $1/2(3)^{1/2} - nr_0^2 \le u \le 2/3\,(3)^{1/2} - nr_0^2$. The current is given by

$$\frac{I}{\pi\,nr_0\,FDc} = \frac{(3)^{1/2}}{\pi u} \left\{ \left[2(3)^{1/2}\left[u+nr_0^2\right] - 1 \right]^{1/2} \right. \qquad [17.2]$$

$$\left. + 2(3)^{1/2}\left[u+nr_0^2\right]\left[\frac{\pi}{6} - \arctan\right]\left[2(3)^{1/2}\left[u+nr_0^2\right]-1\right]^{1/2} \right\}$$

(iii) $r_d > d/(3)^{1/2}$, total coverage of microelectrode array.

For times such that $u > 2/3(3)^{1/2} - nr_0^2$, the microelectrode array is totally covered by diffusion zones. The current is given by

$$\frac{I}{\pi\,nr_0\,FDc} = \frac{1}{\pi u} \qquad [17.3]$$

(c) Random array

Here we consider microdisks randomly distributed on an insulating plane (as would presumably result from imbedding carbon fibers into a polymer). If they are not too closely packed, they will be uniformly distributed, and intermicroelectrode distances can be estimated from the nearest neighbor distribution

$$p(d) = 2\pi \, dn \, \exp(-\pi d^2 n) \tag{18}$$

It can be shown that for such a distribution of distances,

$$n = 1/4 \, (d)^2 \tag{19}$$

the statistical overlap of diffusion zones arising from a uniform distribution of microelectrodes may be treated through Avrami theorem. The fractional area of the ensemble covered by diffusion zones would then be

$$S = 1 - \exp\left[-\frac{\pi r_0 (\pi Dt)^{1/2} + \pi n r_0^2}{4(d)^2}\right] \tag{20}$$

$$S = 1 - \exp\left[-\pi n r_0 (\pi Dt)^{1/2} - \pi n r_0^2\right]$$

The current density to the ensemble is then:

$$\frac{I}{\pi n r_0 \, FDc} = \frac{1}{\pi u}\left\{1 - \exp\left[-\pi\left(u + n r_0^2\right)\right]\right\} \tag{21}$$

The short times expansion of [21] is identical to [15.1] and [17.1], corresponding to diffusion to noninteracting microelectrodes. The long times limit tends to $1/\pi u$, as for the square and hexagonal lattices, equations [15.3] and [17.3]. The transition time $\tau_{s \to p}$, for the spherical-to-planar transition, occurs when the argument of the exponential term in equation [21] becomes of the order of unity. Thus,

$$\tau_{s \to p} = \frac{\left(1 - n\pi r_0^2\right)^2}{\pi D \left(\pi n r_0\right)^2} \qquad [22]$$

which can also be expressed as

$$\tau_{s \to p} = \frac{\theta^2}{\pi D (P/2)^2} \qquad [23]$$

where θ is the electroinactive fractional area and P is the sum of perimeters of the inlaid microelectrodes in the ensemble.

Attilla Szabo contributed the following on his work.

Hemicylinder, Band, Disk and Ring Electrodes

Szabo et al (1) have considered the current at hemicylinder electrodes based on the heat flux models developed earlier (2) for cylinders of infinite length. If the length ℓ of a cylinder electrode of radius a is assumed to be large enough so that the flux at the ends of the cylinder is small compared to the total flux, then one obtains for the current (assuming then the area of the cylinder is $2\pi a\ell$) to a hemicylinder

$$i = nFDC_o \ell \pi f(\theta) \quad [1]$$

where $\theta = Dt/a^2$ and $f(\theta)$, the flux, is given by

$$f(\theta) = \frac{4}{\pi^2} \int_0^\infty \frac{e^{-\theta x^2} dx}{x\left[J_o(x)^2 + Y_o(x)^2\right]} \quad [2]$$

where $J_o(x)$ and $Y_o(x)$ are Bessel functions of the first and second kind, respectively, of order zero. This integral has been evaluated numerically (3). The limiting values

$$f(\theta) = \frac{1}{(\pi\theta)^{1/2}} + \frac{1}{2} - \frac{1}{4(\pi)^{1/2}} \theta^{1/2}$$

$$+ \frac{\theta}{8} - \cdots \quad \theta \to 0 \quad [3]$$

and $$f(\theta) = \frac{2}{\ln(4\theta e^{-2\gamma})} - \frac{2\gamma}{\left[\ln(4\theta e^{-2\gamma})\right]^2}$$

$$- \frac{\left[\frac{\pi^2}{3} - 2\gamma^2\right]}{\left[\ln(4\theta e^{-2\gamma})\right]^3} + \cdots \quad \theta \to \infty \quad [4]$$

where $\gamma = 0.5772156$ have been given (3). These workers find that the approximate form

$$f(\theta) = \frac{e^{(-(\pi\theta)^{1/2})/10}}{(\pi\theta)^{1/2}} + \frac{1}{\ln[(4e^{-\gamma}\theta)^{1/2} + e^{5/3}]} \qquad [5]$$

holds to within 1.3% of the numerically evaluated integral [2] for all times.

The flat band electrode response was approximated by another procedure. At short times, the response (neglecting edge effects at the ends) for a band of width w is obtained using Oldham's approach (4) as

$$i(t) = nFDC_o \ell \left[\frac{1}{(\pi\tau)^{1/2}} + 1 \right] \quad , \; \tau \to 0 \qquad [6]$$

where $\tau = Dt/w^2$.

For long times, using a heuristic analysis the authors find that the current in Laplace space for long times (small s) is given by

$$\overline{i(s)} = \frac{-2\pi nFDC_o \ell}{s \ln\left[\dfrac{sw^2 e^{2\gamma}}{2^6 D}\right]} \qquad [7]$$

which can be compared to the asymptotic long time limit of the current to a hemicylinder, from which it was determined that the currents are the same value when it is assumed that the radius of the hemicylinder is equal to one fourth of the width of the flat band geometry. The authors present arguments that this latter point is probably exact. Further, based on a simulation analysis, the above correspondence, and equations [5] and [6], the band response may be approximated for all times by

$$\frac{i(t)}{nFDC_0 \ell} = \begin{cases} \dfrac{1}{(\pi\tau)^{1/2}} + 1, & \tau < 2/5 \\[2ex] \dfrac{\pi e^{-2/5(\pi\tau)^{1/2}}}{4(\pi\tau)^{1/2}} + \dfrac{\pi}{\left[\ln(64e^{-\gamma}\tau)^{1/2} + e^{5/3}\right]}, & \tau > 2/5 \end{cases} \quad [8]$$

Compared to numerical simulations, it was shown that [8] was accurate to 1.3%. Optimization of some of the empirical parameters leads to more accurate expressions.

For a disk electrode of radius b lying on an inert support, the diffusion limited current can be written as

$$i(t) = 4nFDC_0 b f(\tau) \quad [9]$$

with $\tau = 4Dt/b^2$. Oldham (4) and Aoki and Osteryoung (5) determined the first two terms of the short time expansion of $f(\tau)$, while Shoup and Szabo (6) evaluated the first three terms in the long time expansion using the Wiener-Hopf method (5). Shoup and Szabo (6) found that the empirical expression

$$f(\tau) = A + B\tau^{-1/2} + C \exp(-D\tau^{-1/2}) \quad [10]$$

where

$$A = \pi/4 \qquad [11a]$$

$$B = (\pi)^{1/2}/2 \qquad [11b]$$

$$C = 1 - \pi/4 \qquad [11c]$$

$$D = \frac{\left[\frac{(\pi)^{1/2}}{2} - 4\pi^{-3/2}\right]}{\left[1 - \pi/4\right]} \qquad [11d]$$

reproduces the first two terms in both the exact short and long time expansions and is accurate to 0.6% for *all* times.

Szabo has recently investigated the chronoamperometric response to a ring electrode in an insulator (7). Assuming uniform flux across the ring and requiring the usual boundary condition to hold on the average over the surface of the electrode, Szabo finds, in the Laplace plane,

$$\overline{i(s)} = \frac{nFD\pi(b^2-a^2)(k_f C_O^* - k_b C_R^*)}{s\left[D + 2(k_f + k_b)F(s)/(b^2-a^2)\right]} \qquad [12]$$

with

$$F(s) = \int_0^\infty \frac{[bJ_1(\lambda b) - aJ_1(\lambda a)]^2}{\lambda(\lambda^2 + s/D)^{1/2}} d\lambda \qquad [13]$$

where a and b are the inner and outer radii of the ring, respectively, J_1 is a Bessel function of the first kind of order 1, D is the diffusion coefficient, assumed the same for the oxidized and reduced forms, the C_i^* are the bulk concentration values for the oxidized (i=0) and reduced (i=R) species, λ is a dummy integration variable, and k_f and k_b are the forward and reverse heterogeneous rate constants respectively for the reaction

$$0 + ne^- \underset{k_b}{\overset{k_f}{\rightleftarrows}} R. \qquad [14]$$

For long times, he obtains

$$\frac{i(t)(k_f + k_b)}{nFD\left[k_f C_O^* - k_b C_R^*\right]} = \ell\left[1 + \ell/(4\pi^3 Dt)^{1/2}\right], \quad t \to \infty \qquad [15]$$

where

$$\frac{1}{\ell} = \frac{1}{\ell_o} + \frac{D}{A(k_f + k_b)} \qquad [16]$$

where A is the area of the ring = $\pi(b^2 - a^2)$, and

$$\ell_o = \frac{3\pi^2 (b^2 - a^2)^2}{8(b^3 + a^3) - 4(a+b)[(a^2 + b^2)E(k) - (b-a)^2 K(k)]} \qquad [17a]$$

$$= \frac{\pi^2(a+b)}{\ln\left[4e^{3/2}\left(\frac{b+a}{b-a}\right)\right]}, \quad \frac{(b-a)}{b} \ll 1 \qquad [17b]$$

where $k^2 = 4ab/(a+b)^2$ and $E(k)$, $K(k)$ are complete elliptic integrals. The noteworthy feature of equation [15] is that ℓ determines not only the steady state current but also the coefficient of the $t^{-1/2}$ term (i.e., the way the steady state is approached). The exact value of ℓ_o for thin rings is (8,9)

$$\ell_o = \frac{\pi^2(a+b)}{\ln\left[16\left(\frac{b+a}{b-a}\right)\right]} \quad , \quad \frac{(b-a)}{b} \ll 1 \qquad [18]$$

For $1.02 < b/a < 1.25$, the approximate ℓ's in equation [17b] are within 2.5% - 1.5% of the exact values calculated from equation [18] and improve as the ring becomes thinner. For rings of arbitrary thickness, Szabo (7) devised the empirical formula

$$\ell_o = \frac{\pi^2(a+b)}{\ln[32a/(b-a) + \exp(\pi^2/4)]} \qquad [19]$$

which reproduces the available exact values of ℓ_o, obtained numerically, to within 0.2%. For short times, the exact behavior of the diffusion limited current obtained using Oldham's (4) work, is

$$\frac{i(t)}{nFDC_o^*} = \frac{A}{(\pi Dt)^{1/2}} + \pi(a+b) + \cdots \qquad t \to 0 \qquad [20]$$

Based on a variety of considerations, Szabo (7) conjectured that the long-time behavior of the diffusion-limited current at *any* microelectrode with a closed surface lying on an effectively infinite planar support has the form

$$i(t) = nFDC_o^*\ell_o\left[1+\ell_o/(4\pi^3 Dt)^{1/2}\right] \qquad t \to \infty \qquad [21]$$

where ℓ_o has dimensions of length and depends on the geometry of the electrode. Available exact results for a disk (6) and hemispherical electrodes support this conjecture (i.e. $\ell_o = 4b$ for a disk of radius b and $\ell_o = 2\pi R$ for a hemisphere of radius R). It was suggested that for a ring, the essentially exact current at long times is given by equation [21] with ℓ_o given by equation [19].

References

1. A. Szabo, D.K. Cope, D.E. Tallman, P.M. Kovach and R.M. Wightman, J. Electroanal. Chem., 217 (1987) 417.
2. H.S. Carslow and J.C. Jaeger, "Conduction of Heat in Solids", Oxford Press, 1959, pp. 334-336 and references therein.
3. J.C. Jaeger and R. Clarke, Proc. Roy. Soc. Edin., A 61 (1942) 233.
4. K.B. Oldham, J. Electroanal. Chem., 122 (1982) 1.
5. K. Aoki and J. Osteryoung, J. Electroanal. Chem., 122 (1981) 19.
6. D. Shoup and A. Szabo, J. Electroanal. Chem., 140 (1982) 237.
7. A. Szabo, J. Phys. Chem., 91 (1987) 3108.
8. W.R. Smythe, J. Appl. Phys., 22 (1951) 1499.
9. I.S. Symanski and S. Bruckenstein, Extended Abstract, 165th Meeting of the Electrochemical Society.

The following is a summary of the lecture presented by R. Mark Wightman.

The temporal response of Nafion-coated, microvoltammetric electrodes to rectangular concentration changes has been evaluated. These devices, with micrometer tip dimensions, have been designed for use as *in vivo* probes of electroactive neurotransmitters such as dopamine. The permeation of dopamine at electrodes with Nafion films thinner than 200nm is sufficiently fast that distortion of the concentration pulse is not apparent. The permeation rate of $Ru(bpy)_3^{2+}$ into these films varies with the applied potential waveform. At constant applied potential the measured permeation rate increases with repetitive exposures. However, with fast-scan cyclic voltammetry ($200V \cdot s^{-1}$) repeated at 1s intervals the permeation time for $Ru(bpy)_3^{2+}$ remains relatively constant with repetitive exposures. Physical diffusion through these films predominates for this complex when sampled in this manner. The thickness of each individual film can be determined from the permeation rates. The diffusion coefficient for dopamine in these thin Nafion films is found to decrease with a decrease in sodium ion concentration in solution. Voltammograms of dopamine in Nafion films are similar to those in solution except at high loading of the film where the buffer capacity can be exceeded. The electrodes are found to respond rapidly to the stimulated secretion of dopamine in the brain of an anesthetized rat.

Marek Wojciechowski contributed the following regarding his view on some of the directions that should be taken in electrochemical analysis with ultramicroelectrodes.

The Utah Meeting on Ultramicroelectrodes was a very productive and stimulating meeting which summarized the latest achievements and formulated future goals of research in the field of ultramicroelectrodes. Particularly impressive have been recent advances in methods of fabrication of microelectrode devices, and in methodology of measurements with microelectrodes. Now this technology and methodology deserve broader implementation in solving real chemical problems and in analysis. Several very attractive possibilities in these areas were addressed during the meeting.

My comments below are related to one area of ultramicroelectrode research (a focus of my own research interests); the utilization of ultramicroelectrodes in electroanalysis.

1. There are several unique properties of ultramicroelectrodes which make them an attractive tool in modern electroanalysis; the most important are:

- small size—analytical experiment can be performed directly in microdimensional systems like biological cells, pores, or microvolume samples;

- greater analytical sensitivity than electrodes of conventional size—this attribute is related to large current densities resulting from enhanced mass transport due to nonplanar diffusion at ultramicroelectrodes; improved ratios of faradaic current to nonfaradaic current are also observed;

- fast response time;

- diminished iR problems;

- relatively easy and cheap to make.

2. Advanced voltammetric techniques, such as ultrafast cyclic and square wave voltammetries, begin to play a very important role in measurements with ultramicroelectrodes. Considering the attributes offered by the combination of these techniques with ultramicroelectrodes, it seems certain that this combination will dominate the electroanalytical methodology in the future. The ultrafast (kV/s sweep-rates) cyclic or linear scan voltammetry, for example, offers very low detection limits at extremely short times required for a single scan. The fast scanning capability of the technique can be utilized for improving the signal-to-noise ratio through averaging of multiple scans. Subtraction of the background curve is necessary, however, in order to improve the analytical signal.

Another voltammetric technique which seems even better suited for measurements with ultramicroelectrodes is square wave voltammetry. This technique is very fast and sensitive, but it also offers rejection of background currents. Therefore, in contrast with linear scan or cyclic voltammetry, there is no need for background subtraction. One of the unique features offered by square wave voltammetry at ultramicroelectrodes is a peak-shape response whose shape is invariant to electrode geometry and size, time scale of experiment, and mechanism of mass transport to the electrode surface. This latter attribute makes the technique insensitive to convective mass transport, even for larger electrodes, and suggests very interesting applications for measurements in hydrodynamic systems. Our group is pursuing work devoted to characterization of square wave voltammetry at both macro- and microelectrodes under forced convection. Potential applications of hydrodynamic square wave voltammetry include: continuous monitoring in flowing streams; 3-D electrochemical detection in HPLC and flow injection analysis; studies where it is necessary or desirable to use turbulent solutions, and where the convective mass transport may be difficult to characterize or to control; measurements in convective microenvironments, such as membrane pores, etc..

We are also working on the application of square wave voltammetry with ultramicroelectrodes in anodic (transition metal ions) and cathodic (thio-organics) stripping analysis.

Another very important aspect of square wave voltammetry is its use for assessment of fabrication quality of ultramicroelectrodes.

3. Minimal iR offered by microelectrodes should allow measurements in highly resistive media, such as poorly conductive solvents or solutions without supporting electrolyte. In trace electroanalysis, for example, the absence of supporting electrolyte can eliminate problems caused by impurities introduced with the electrolyte. Also, the absence of supporting electrolyte will significantly extend the range of potentials accessible for electrochemical measurements, since "the potential window" will then be bracketed exclusively by depolarization of the working electrode involving the solvent and/or the electrode material. Therefore, electrochemical processes that require extremely negative or positive potentials could be studied and, possibly, utilized analytically. This application of microelectrodes, however, still requires thorough experimental verification. Of particular importance among the problems that must be addressed in microelectrode electrochemistry with little or no electrolyte, are: double-layer effects; involvement of migration; effect of species, especially ionic, produced at the working and counter electrodes; contaminations from the cell material, and electrode materials and supports.

4. The application of ultramicroelectrodes in anodic stripping voltammetry and analysis has been demonstrated as another very promising and attractive field. In our laboratory we are interested in developing reliable procedures for plating mercury "films" on microelectrodes made from graphite fibers and metallic wires. We are currently examining different approaches in electrochemical pretreatment of cylindrical graphite fiber microelectrodes, with regard to:

- reproducibility of film formation (nucleation and growth of mercury nuclei),

- anodic dissolution of mercury films (the recovery problem),

- chemical and mechanical durability of the film, especially in the presence of dissolved oxygen,

- stability of amalgams formed *in situ*.

Ultramicroelectrodes having nonflat surfaces, such as our cylindrical microelectrodes, can not be polished mechanically. Therefore, a reproducible deposition of thin mercury films presents even more difficult problem than in case of polished microelectrodes. We are investigating the film formation process on differently pretreated graphite fibers, using double potential step chronocoulometry as well as pulse voltammetric techniques. Fibers preplated with Ag, Au or other metals will also be examined.

5. A very promising application of ultramicroelectrodes, although still requiring a great deal of work before it can be implemented in practice, is the analysis of microvolume samples. This topic has been investigated in our laboratory. We are currently testing a new, two-electrode microelectrochemical cell of our own design. The cell consists of a 1mm long graphite

fiber electrode sealed with epoxy in a glass capillary which is
mounted in a stainless-steel needle. The needle serves as a
cell body and as a reference electrode. A 10µL volume of a test
solution can be aspirated into and dispensed from this cell by
means of an Eppendorf pipet attached to the needle. With a
different aspirating device, it should be possible to work with
much smaller volumes. We are currently testing the performance
of this micro-cell. For that purpose square wave anodic
stripping voltammetry of lead and cadmium is being used, with *in
situ* plated mercury film on the graphite fiber electrode, and in
air-saturated samples. The depletion of oxygen during amalgam
deposition, combined with the fast-scanning capabilities of the
technique, allows us to carry out the entire experiment in the
presence of dissolved oxygen.

6. With regard to microelectrode technology, intensified
research efforts are needed in the following areas:

- Fabrication of sub-micron electrodes that would not require
  mechanical treatments for reliable and reproducible
  functioning.

- Development of ultramicroelectrodes with *overall* dimensions,
  including the support, on the order of 1µm. These
  microscopic devices are prerequisite for advancement of
  electrochemistry in the basic and applied studies of
  microenvironments such as corrosion pits, membrane pores or
  single biological cells, for example.

- Development of reference microelectrodes that would perform
  in a reliable fashion in voltammetric or coulometric
  measurements employing two-electrode configurations. So
  far, reference electrodes have been somewhat neglected in
  the ultramicroelectrode research. However, it is obvious
  that reference electrodes of similar dimensions as working
  ultramicroelectrodes, are necessary for successful
  miniaturization of electrochemical probes.

- Development of procedures that could be readily adoptable for commercial production of microelectrodes. Microelectrodes must be developed which will be ready and easy in use, reliable for some time without treatment, inexpensive (possibly disposable), available in some selected geometries (disk, single or array, and cylinder, for example) for the most common electrode materials such as Pt, Au, Ag and graphite. It seems clear that the field of ultramicroelectrode research has reached the state where much more could be achieved, especially in terms of electrochemical and electroanalytical applications, if there were microelectrodes available commercially.

- Characterization and utilization of microelectrodes having relatively large total electrode area. These are single microelectrodes of cylindrical or linear geometry, or ensembles of interconnected, regularly or randomly distributed microelectrodes whose shape and size may vary. These electrodes exhibit microelectrode characteristics with respect to the nonplanar diffusion control of the mass-transport, while generating the currents that are sufficiently large to be measured with conventional instrumentation. No need for a sophisticated, low-current potentiostat makes this type of microelectrode very attractive for the work in media where the iR drop does not present a serious problem.

6

Selected Bibliographies

James L. Anderson

J.L. Anderson and S. Moldoveanu. "Numerical Simulation of Convective Diffusion at a Rectangular Channel Electrode," J. Electroanal. Chem., 179 (1984) 107-117.

S. Moldoveanu, G. Handler, and J.L. Anderson. "On Convective Mass Transfer in Laminar Flow between Two Parallel Electrodes in a Rectangular Channel," J. Electroanal. Chem., 179 (1984) 119-130.

S. Moldoveanu and J.L. Anderson. "Numerical Simulation of Convective Diffusion at a Microarray Channel Electrode," J. Electroanal. Chem., 185 (1985) 239-252.

J.L. Anderson, K.K. Whiten, J.D. Brewster, T.Y. Ou, and W.K. Nonidez. "Microarray Electrochemical Flow Detectors at High Applied Potentials and Liquid Chromatography with Electrochemical Detection of Carbamate Pesticides in River Water," Anal. Chem., 57 (1985) 1366-1373.

J.L. Anderson, T.Y. Ou, and S. Moldoveanu. "Hydrodynamic Voltammetry at an Interdigitated Electrode Array in a Flow Channel. I. Numerical Simulation," J. Electroanal. Chem., 196 (1985) 213-226.

F. Belal and J.L. Anderson. "Flow Injection Analysis of Three N-Substituted Phenothiazine Drugs with Amperometric Detection at a Carbon Fibre Array Electrode," Analyst, 110 (1985) 1493-1496.

F. Belal and J.L. Anderson. "Flow Injection Determination of Ergonovine Maleate with Amperometric Detection at the Kel-F-Graphite Composite Electrode," Talanta, 33 (1986) 448-450.

L.E. Fosdick and J.L. Anderson. "Optimization of Microelectrode Array Geometry in a Rectangular Flow Channel Detector," Anal. Chem., 58 (1986) 2481-2485.

L.E. Fosdick, J.L. Anderson, T.A. Baginski, and R.C. Jaeger. "Amperometric Response of Microlithographically Fabricated Microelectrode Array Flow Sensors in a Thin-Layer Channel," Anal. Chem., 58 (1986) 2750-2756.

Q.G. von Nehring, J.W. Hightower, and J.L. Anderson. "Liquid Chromatography with Electrochemical Detection and Coulometric Investigations of Carbamate and Urea Pesticides," Anal. Chem., 58 (1986) 2777-2781.

Koichi Aoki

K. Aoki and J. Osteryoung. "Diffusion-Controlled Current at the Stationary Finite Disk Electrode: Theory," J. Electroanal. Chem., 122 (1981) 19-36.

K. Aoki and J. Osteryoung. "Diffusion-Controlled Current at Stationary Finite Disk Electrode: Experiment," J. Electroanal. Chem., 125 (1981) 315-320.

K. Aoki and J. Osteryoung. "Formulation of the Diffusion-Controlled Current at Very Small Stationary Disk Electrodes," J. Electroanal. Chem., 160 (1984) 335-340.

K. Aoki, K. Akimoto, K. Tokuda, H. Matsuda and J. Osteryoung. "Linear Sweep Voltammetry at Very Small Stationary Disk Electrodes," J. Electroanal. Chem., 171 (1984) 219-230.

K. Aoki, H. Honda, K. Tokuda and H. Matsuda. "Voltammetry at Microcylinder Electrodes. Part I. Linear Sweep Voltammetry," J. Electroanal. Chem., 182 (1985) 267-279.

K. Aoki, K. Aokimoto, K. Tokuda, H. Matsuda and J. Osteryoung. "Chronopotentiometry at Very Small Stationary Disk Electrodes," J. Electroanal. Chem., 182 (1985) 281-294.

K. Aoki, K. Honda, K. Tokuda and H. Matsuda. "Voltammetry at Microcylinder Electrodes. Part II. Chronoamperometry," J. Electroanal. Chem., 186 (1985) 79-86.

J. O'Dea, M. Wojciechowski, J. Osteryoung and K. Aoki. "Square Wave Voltammetry at Electrodes Having a Small Dimension," Anal. Chem., 57 (1985) 954-955.

K. Aoki, H. Honda, K. Tokuda and H. Matsuda. "Voltammetry at Microcylinder Electrodes. Part III. Chronopotentiometry," J. Electroanal. Chem., 195 (1985) 51-62.

S. Sujaritvanichpong, K. Aoki, K. Tokuda and H. Matsuda. "Electrochemical Behavior of Dopamine at Carbon Fiber Electrodes," J. Electroanal. Chem., 198 (1986) 195-203.

S. Sujaritvanichpong, K. Aoki, K. Tokuda and H. Matsuda. "Voltammetry at Microcylinder Electrodes. Part IV. Normal Pulse and Differential Pulse Voltammetry," J. Electroanal. Chem., 199 (1986) 271-283.

D. Whelan, J.J. O'Dea, J. Osteryoung and K. Aoki. "Square Wave Voltammetry at Small Disk Electrodes: Theory and Experiment," J. Electroanal. Chem., 202 (1986) 23-36.

K. Aoki, K. Tokuda and H. Matsuda. "Voltammetry at Microcylinder Electrodes. Part V. Pulse Voltammetric Current-Voltage Curves for Quasi-Reversible and Totally Irreversible Electrode Reactions," J. Electroanal. Chem., 206 (1986) 47-56.

D. Kagaku, K. Aoki, K. Tokuda and H. Matsuda. "Theory of Chronoamperometric Curves for a Short Time at Microband Electrodes," J. Electrochem. Soc. Jpn., 54 (1986) 1010-1017.

K. Aoki, K. Tokuda and H. Matsuda. "Hydrodynamic Voltammetry at Channel Electrodes. Part IX. Edge Effects at Rectangular Channel Flow Microelectrodes," J. Electroanal. Chem., 217 (1987) 33-48.

K. Aoki, K. Tokuda and H. Matsuda. "Theory of Chronoamperometric Curves at Microband Electrodes," J. Electroanal. Chem., in press.

K. Aoki, K. Tokuda and H. Matsuda. "Derivation of Approximate Equation for Chronoamperometric Curves at Microband Electrodes and the Experimental Verification," J. Electroanal. Chem., submitted.

K. Aoki, K. Tokuda and H. Matsuda. "Theory of Stationary Current-Potential Curves at Microdisk Electrodes for Quasi-Reversible and Totally Irreversible Electrode Reactions," J. Electroanal. Chem., submitted.

K. Aoki and K. Tokuda. "Linear Sweep Voltammetry at Microband Electrodes," J. Electroanal. Chem., submitted.

Alan M. Bond

W. Thormann, P. van den Bosch and A.M. Bond. "Voltammetry at Linear Gold and Platinum Microelectrode Arrays Produced by Lithographic Techniques," Anal. Chem., 57 (1985) 2764-2770.

J.W. Bixler, A.M. Bond, P.A. Lay, W. Thormann, P. van den Bosch, M. Fleischmann and S. Pons. "Instrumental Configurations for the Determination of Sub-Micromolar Concentrations of Electroactive Species with Carbon, Gold and Platinum Microdisk Electrodes in Static and Flow-Through Cells," Analytica Chimica Acta, 187 (1986) 67-77.

A.M. Bond, T.L.E. Henderson and W. Thormann. "Theory and Experimental Characterization of Linear Gold Microelectrodes with Submicrometer Thickness," J. Phys. Chem., 90 (1986) 2911-2917.

J.W. Bixler and A.M. Bond. "Amperometric Detection of Picomole Samples in a Microdisk Electrochemical Flow-Jet Cell with Dilute Supporting Electrolyte," Analytical Chemistry, 58 (1986) 2859-2863.

W. Thormann and A.M. Bond. "Application of Transient Electrochemical Techniques to Inlaid Ultramicroelectrodes," J. Electroanal. Chem., 218 (1987) 187-196.

A.M. Bond and F.G. Thomas. "Studies of Adsorption Processes in the Absence of Added Electrolyte: Phase Changes in Coumarin Adsorbed at Conventional and Micro Mercury Electrodes," private communication and to be published.

A.M. Bond and T.F. Mann. "Voltammetric Measurements without Ohmic and Other Forms of Distortion in Aromatic Hydrocarbon Solvents," private communication and to be published.

Martin Fleischmann

J. Cassidy, S. B. Khoo, S. Pons and M. Fleischmann. "Electrochemistry at Very High Potentials: The Use of Ultramicroelectrodes in the Anodic Oxidation of Short Chain Alkanes", J. Phys. Chem., 89 (1985) 3933.

M. Fleischmann, J. Ghoroghchian and S. Pons. "Electrochemical Behavior of Dispersions of Spherical Ultramicroelectrodes. Part 1. Theoretical Considerations," J. Phys. Chem., 89 (1985) 5530.

M. Fleischmann, S. Bandyopadhyay and S. Pons. "The Behavior of Microring Electrodes," J. Phys. Chem., 89 (1985) 5537.

A. M. Bond, M. Fleischmann, S. B. Khoo, S. Pons and J. Robinson. "The Construction and Behavior of Ultramicroelectrodes: Investigations of Novel Electrochemical Systems," Ind. J. Tech., 24 (1986) 492.

J. W. Bixler, A. M. Bond, D. A. Lay, M. Fleischmann and S. Pons. "Instrumental Configurations for the Determination of Submicromolar Concentrations of Electroactive Species with Carbon, Gold, and Platinum Microdisk Electrodes in Static and Flow Through Cells," Anal. Chim. Acta, 187 (1986) 67.

M. Fleischmann, J. Ghoroghchian, D. R. Rolison and S. Pons. "The Electrochemical Behavior of Dispersions of Spherical Ultramicroelectrodes," J. Phys. Chem., 90 (1986) 6392.

T. Dibble, S. Bandyopadhyay, J. Ghoroghchian, J. J. Smith, F. Sarfarazi, M. Fleischmann and S. Pons. "Electrochemistry at Very High Potentials. Oxidation of the Rare Gases and Other Gases in Non-Aqueous Solvents at Ultramicroelectrodes," J. Phys. Chem., 90 (1986) 5275.

A. Russell, K. Repka, T. Dibble, J. Ghoroghchian, J. J. Smith, M. Fleischmann, C. H. Pitt and S. Pons. "Determination of Electrochemical Heterogeneous Electron Transfer Rates from Steady State Measurements at Ultramicroelectrodes," Analytical Chemistry, 58 (1986) 2961.

J. Ghoroghchian, F. Sarfarazi, T. Dibble, J. Cassidy, J. J. Smith, A. Russell, M. Fleischmann and S. Pons. "Electrochemistry in the Gas Phase. Use of Ultramicroelectrodes for the Analysis of Electroactive Species in Gas Mixtures," Analytical Chemistry, 58 (1986) 2278.

M. Fleischmann, N. Garrard, J. Daschbach and S. Pons.
"Electrochemical Kinetics Using Single Electron Counting,"
Journal of Electroanalytical Chemistry, in preparation.

D.R. Rolison, R.J. Nowak, S. Pons, M. Fleischmann and J.
Ghoroghchian. "Alumina and Zeolyte Supported Ultramicroelectrodes," J. Phys. Chem., in preparation.

M. Fleischmann and S. Pons. "The Behavior of Microring and
Microdisk Electrodes," J. Electroanal. Chem., $\underline{222}$ (1987) 107.

H.Y.S. Lui, M. Phil. Thesis, University of Southampton, (1975).*

D. Swan, Ph.D. Thesis, University of Southampton, (1981).*

F. Lasserre, Ph.D. Thesis, University of Southampton, (1983).*

M. Fleischmann, F. Lasserre, J. Robinson, and D. Swan. "An
Investigation of Coupled Homogeneous Reactions Using
Microelectrodes. Part 1. EC' and CE Reactions," J. Electrochem.
Soc., $\underline{177}$ (1984) 97.

P. Bindra, A.P. Brown, M. Fleischmann and D. Pletcher. "The
Determination of the Kinetics of Very Fast Electrode Reactions
by Means of a Quasi Steady State Method. The Mercurous Ion/
Mercury System. Part I. Theory.," J. Electroanal. Chem., $\underline{58}$
(1975) 31.

P. Bindra, A.P. Brown, M. Fleischmann and D. Pletcher. "The
Determination of the Kinetics of Very Fast Electrode Reactions
by Means of a Quasi Steady State Method. The Mercurous Ion/
Mercury System. Part II. Experimental.," J. Electroanal.
Chem., $\underline{58}$ (1975) 39.

M. Fleischmann and H.R. Thirsk in Advances in Electrochemistry
and Electrochemical Engineering, Ed. P. Delahay, Interscience,
New York and London, Volume 3, 123, (1963).

M. Fleischmann, F. Goodridge and C.J.H. King, Brit. Pat. Appl.
16765 (1974).

M. Fleischmann, F. Lasserre and J. Robinson. An Investigation
of Coupled Homogeneous Reactions Using Microelectrodes. Part
II. ECE and DISP1 Reaction," J. Electroanal. Chem., $\underline{177}$ (1984)
815.

A.M. Bond, M. Fleischmann and J. Robinson. "Voltammetric Measurements Using Microelectrodes in Highly Dilute Electrolyte Solutions: Theoretical Considerations," J. Electroanal. Chem., 172 (1984) 11.

A.M. Bond, M. Fleischmann and J. Robinson. "The Use of Platinum Microelectrodes for Electrochemical Investigations in Low Temperature Glasses of Nonaqueous Solvents," J. Electroanal. Chem., 180 (1984) 257.

A.M. Bond, M. Fleischmann and J. Robinson. "Electrochemistry in Organic Solvents without Supporting Electrolyte Using Platinum Microelectrodes," J. Electroanal. Chem., 168 (1984) 299.

P. Bindra, M. Fleischmann, J.W. Oldfield and D. Singleton. "Nucleation," Faraday Disc. Chem. Soc., 56 (1973) 180.

M. Fleischmann, C. Gabrielli, M. Labrum and A. Sattar. "The Measurement and Interpretation of Stochastic Effects in Electrochemistry and Bioelectrochemistry," Surf. Sci., 101 (1980) 583.

E. Budevski, M. Fleischmann, C. Gabrielli and M. Labrum. "Statistical Analysis of the 2-D Nucleation and Electrocrystallization of Silver," Electrochim. Acta, 28 (1983) 925.

S. Pons, M. Fleischmann, J. Pons and J. Daschbach. "The Behavior of the Mercury Ultramicroelectrode," J. Electroanal. Chem., submitted.

*Theses directed by M. Fleischmann

Royce W. Murray

A. G. Ewing, B. Feldman, and R. W. Murray. "Polymer Coated Microelectrodes and Twin Electrode Thin Layer Cells Applied to Electron Transfer Cross-Reaction and Permeability Rates," J. Electroanal. Chem., 172 (1984) 145.

P. Pickup, W. Kutner, C. R. Leidner and R. W. Murray. "Redox Conduction in Single and Bilayer Films of Redox Polymer," J. Am. Chem. Soc., 106 (1984) 1991.

P. Pickup and R. W. Murray. "Redox Conduction in Mixed Valent Polymers," J. Am. Chem. Soc., 105 (1983) 4510.

A. G. Ewing, B. J. Feldman, and R. W. Murray. "Permeation of Neutral, Cationic, and Anionic Electrode Reactants Through a PolyCationic Polymer Film as A Function of Electrolyte Concentration," J. Phys. Chem., <u>89</u> (1985) 1263.

J. C. Jernigan, C. E. D. Chidsey, and R. W. Murray. "Electrochemistry of Polymers Not Immersed in Solvent: Electron Transfer on an Ion Budget," J. Am. Chem. Soc., <u>107</u> (1985) 2824.

B.J. Feldman, A.E. Ewing, and R.W. Murray. "Electron Transfer Kinetics at Redox Polymer Film/Solution Interfaces Using Microelectrodes and Twin Electrode Thin Layer Cells," J. Electroanal. Chem., <u>194</u> (1985) 63.

C.E.D. Chidsey and R.W. Murray. "Electroactive Polymers and Macromolecular Electronics", Science, <u>231</u> (1986) 25.

C.E.D. Chidsey, B.J. Feldman, C. Lundgren, and R.W. Murray. "Micrometer-Spaced Platinum Interdigitated Array Electrode: Fabrication, Theory, and Initial Use," Anal. Chem., <u>58</u> (1986) 601.

L. Geng, A.G. Ewing, J.C. Jernigan, and R.W. Murray. "Electrochemistry and Electroactive Polymer Films in Low Dielectric Solvents: Toluene and Heptane," Anal. Chem., <u>58</u> (1986) 852.

C.E.D. Chidsey and R.W. Murray. "Redox Capacity and DC Electron Conductivity in Electroactive Materials," J. Phys. Chem., <u>90</u> (1986) 1479.

R.A. Reed, L. Geng, and R.W. Murray. "Solid State Voltammetry of Electroactive Couples in Polyethylene Oxide Films on Microelectrodes," J. Electroanal. Chem., <u>185</u> (1986) 208.

L. Geng and R.W. Murray. "Oxidative Microelectrode Voltammetry of Tetraphenylporphyrin and of Copper Tetraphenylporphyrin in Toluene Solvent," Inorg. Chem., <u>25</u> (1986) 3115.

B.J. Feldman and R.W. Murray. "Measurement of Electron Diffusion Coefficients Through Prussian Blue Electroactive Films Electrodeposited on Interdigitated Array Pt Electrodes," Anal. Chem., <u>58</u> (1986) 2844.

J.C. Jernigan and R.W. Murray. "Electron Self Exchanges in an Osmium Polypyridine Redox Polymer in the Absence of Liquid Solvents by Solid State Voltammetry," J. Am. Chem. Soc., in press.

B.J. Feldman and R.W. Murray. "Electron Diffusion in Wet and Dry Prussian Blue Films on Interdigitated Array Electrodes," Inorg. Chem., in press.

L. Geng, R.A. Reed, and R.W. Murray. "Solid State Linear Sweep Voltammetric Current Responses of Thin Films on Microdisk Electrodes. A Probe of Diffusion in Polymer Ion Conductors," J. Phys. Chem., in press.

J.C. Jernigan and R.W. Murray. "Consequences of Restricted Ion Mobility in Electron Transport Through Films of a Polymeric Osmium Poly-Pyridine Complex," J. Phys. Chem., in press.

B.J. Feldman, S.W. Feldberg and R.W. Murray. "An Electrochemical Time of Flight Experiment," submitted for publication.
R.W. Murray. "Chemically Modified Electrodes," Welch Foundation Conference on Advances in Electrochemistry, October 1986, in press.

Janet G. Osteryoung

K. Aoki and J. Osteryoung. "Diffusion-Controlled Current at the Stationary Finite Disk Electrode: Theory," J. Electroanal. Chem., 122 (1981) 19-35.

K. Aoki and J. Osteryoung. "Diffusion-Controlled Current at the Stationary Finite Disk Electrode: Experiment," J. Electroanal. Chem., 125 (1981) 315-320.

T.R. Brumleve and J. Osteryoung. "Spherical Diffusion and Shielding Effects in Reverse Pulse Voltammetry," J. Phys. Chem., 86 (1982) 1794-1801.

T. Hepel and J. Osteryoung. "Chronoamperometric Transients at the Stationary Disk Microelectrode," J. Phys. Chem., 86 (1982) 1406-1411.

T. Hepel, W. Plot and J. Osteryoung. "Correction to Chronoamperometric Transients at the Stationary Disk Microelectrode," J. Phys. Chem., 87 (1983) 1278.

N. Sleszynski, J. Osteryoung and M. Carter. "Arrays of Very Small Electrodes Based on Reticulated Vitreous Carbon," Anal. Chem., 56 (1984) 130-135.

K. Aoki and J. Osteryoung. "Formulation of the Diffusion-Controlled Current at Very Small Stationary Disk Electrodes," J. Electroanal. Chem., **160** (1984) 335-339.

K. Aoki, K. Akimoto, K. Tokuda, H. Matsuda, and J. Osteryoung. "Linear Sweep Voltammetry at Very Small Stationary Disk Electrodes," J. Electroanal. Chem., **171** (1984) 219-230.

K. Aoki, K. Akimoto, K. Tokuda, H. Matsuda, and J. Osteryoung. "Chronopotentiometry at Very Small Stationary Disk Electrodes," J. Electroanal. Chem., **182** (1985) 281-294.

J.J. O'Dea, M. Wojciechowski, J. Osteryoung and K. Aoki. "Square Wave Voltammetry at Electrodes Having a Small Dimension," Anal. Chem., **57** (1985) 954-955.

T. Hepel and J. Osteryoung. "Relaxation Spectrum for Small Passivating Chromium Electrodes," J. Electrochem. Soc., **133** (1986) 757-760.

T. Hepel and J. Osteryoung. "Electrochemical Characterization of Electrodes with Submicron Dimensions," J. Electrochem. Soc., **133** (1986) 752-757.

D. Whelan, J.J. O'Dea, J. Osteryoung and K. Aoki. "Square Wave Voltammetry at Small Disk Electrodes: Theory and Experiment," J. Electroanal. Chem., **202** (1986) 23-36.

J. Golas and J. Osteryoung. "Mercury-coated Carbon Fiber Microelectrodes: Preparation and Some Properties," Anal. Chim. Acta, **181** (1986) 211-218.

K. Aoki, K. Tokuda, H. Matsuda and J. Osteryoung. "Reversible Square Wave Voltammograms: Independent of Electrode Geometry," J. Electroanal. Chem., **207** (1986) 25-39.

J. Golas and J. Osteryoung. "Carbon Fiber Microelectrodes as Substrates for Mercury Films," Anal. Chim. Acta, **186** (1986) 1-9.

J. Golas, Z. Galus and J. Osteryoung. "Iridium-based Small Mercury Electrodes," Anal. Chem., **59** (1987) 386-389.

J. Golas and J. Osteryoung. "Electrodeposition and Stripping of Silver on Single Carbon Fibers," Anal. Chim. Acta, in press.

Z. Galus and J. Osteryoung. "Linear Scan Voltammetry and Chronoamperometry at Small Mercury Film Electrodes," Electrochim. Acta, in press.

Z. Galus, J. Golas and J. Osteryoung. "Determination of Kinetic Parameters from Steady State Microdisk Voltammograms," submitted to J. Phys. Chem.

Stanley Pons

B. Speiser and S. Pons. "Simulation of Edge Effects. 1. Two Dimensional Collocation and Theory for Chronoamperometry," Canadian Journal of Chemistry, 60 (1982) 1352.

B. Speiser and S. Pons. "Simulation of Edge Effects. 2. Theory for Cyclic Voltammetry," Canadian Journal Of Chemistry, 60 (1982) 2463.

B. Speiser and S. Pons. "Simulation of Edge Effects. 3. Application to Chronoamperometric Experiments," Canadian Journal Of Chemistry, 61 (1983) 156.

S. Pons. "Polynomial Approximation Techniques for Differential Equations in Electrochemical Problems," in A. Bard, ed., "Electroanalytical Chemistry," vol. 13. Marcel Dekker, New York, 1984.

J. Cassidy, S. Pons, and B. Speiser. "Simulation of Edge Effects. Part 4. Application to Cyclic Voltammetry," Canadian Journal Of Chemistry, 62 (1984) 716.

J. Cassidy, S. B. Khoo, S. Pons and M. Fleischmann. "Electrochemistry at Very High Potentials: The Use of Ultramicroelectrodes in the Anodic Oxidation of Short Chain Alkanes," J. Phys. Chem., 89 (1985) 3933.

M. Fleischmann, J. Ghoroghchian and S. Pons. "Electrochemical Behavior of Dispersions of Spherical Ultramicroelectrodes. Part 1. Theoretical Considerations," J. Phys. Chem., 89 (1985) 5530.

M. Fleischmann, S. Bandyopadhyay and S. Pons. "The Behavior of Microring Electrodes," J. Phys. Chem., 89 (1985) 5537.

J. Cassidy and S. Pons. "Simulation of Edge Effects in Electroanalytical Experiment by Orthogonal Collocation. Part V. Chronoamperometry at Ultramicroelectrode Ensembles," Canadian J. Chem., 63 (1985) 3577.

A. M. Bond, M. Fleischmann, S. B. Khoo, S. Pons and J. Robinson. "The Construction and Behavior of Ultramicroelectrodes: Investigations of Novel Electrochemical Systems," Ind. J. Tech., 24 (1986) 492.

J. W. Bixler, A. M. Bond, D. A. Lay, M. Fleischmann and S. Pons. "Instrumental Configurations for the Determination of Submicromolar Concentrations of Electroactive Species with Carbon, Gold, and Platinum Microdisk Electrodes in Static and Flow Through Cells," Anal. Chim. Acta, 187 (1986) 67.

J. Cassidy, J. Ghoroghchian, F. Sarfarazi, J. J. Smith and S. Pons. "Simulation of Edge Effects in Electroanalytical Experiments by Orthogonal Collation. Part VI. Cyclic Voltammetry and Ultramicroelectrode Ensembles," Electrochimica Acta, 31 (1986) 629.

J. Cassidy, J. Janata and S. Pons. "Hydrogen Response of Palladium Coated Suspended Gate Field Effect Transistor," Analytical Chemistry, 58 (1986) 1757.

M. Fleischmann, J. Ghoroghchian, D. R. Rolison and S. Pons. "The Electrochemical Behavior of Dispersions of Spherical Ultramicroelectrodes," J. Phys. Chem., 90 (1986) 6392.

T. Dibble, S. Bandyopadhyay, J. Ghoroghchian, J. J. Smith, F. Sarfarazi, M. Fleischmann and S. Pons. "Electrochemistry at Very High Potentials. Oxidation of the Rare Gases and Other Gases in Non-Aqueous Solvents at Ultramicroelectrodes," J. Phys. Chem., 90 (1986) 5275.

A. Russell, K. Repka, T. Dibble, J. Ghoroghchian, J. J. Smith, M. Fleischmann, C. H. Pitt and S. Pons. "Determination of Electrochemical Heterogeneous Electron Transfer Rates from Steady State Measurements at Ultramicroelectrodes," Analytical Chemistry, 58 (1986) 2961.

J. Ghoroghchian, F. Sarfarazi, T. Dibble, J. Cassidy, J. J. Smith, A. Russell, M. Fleischmann and S. Pons. "Electrochemistry in the Gas Phase. Use of Ultramicroelectrodes for the Analysis of Electroactive Species in Gas Mixtures," Analytical Chemistry, 58 (1986) 2278.

M. Fleischmann, N. Garrard, J. Daschbach and S. Pons. "Electrochemical Kinetics Using Single Electron Counting," Journal of Electroanalytical Chemistry, in preparation.

D.R. Rolison, R.J. Nowak, S. Pons, M. Fleischmann and J. Ghoroghchian. "Alumina and Zeolyte Supported Ultramicroelectrodes," J. Phys. Chem., in preparation.

M. Fleischmann and S. Pons. "The Behavior of Microring and Microdisk Electrodes," J. Electroanal. Chem., 222 (1987) 107.

S. Pons, M. Fleischmann, J. Pons and J. Daschbach. "The Behavior of the Mercury Ultramicroelectrode," J. Electroanal. Chem., submitted.

Hannah Reller

H. Reller, E. Kirowa-Eisner and E. Gileadi. "Ensembles of Microelectrodes--A Digital Simulation," J. Electroanal. Chem., 130 (1982) 65-77.

H. Reller, E. Kirowa-Eisner and E. Gileadi. "Ensembles of Microelectrodes: Digital Simulation by the Two-Dimensional Expanding Grid Method. Cyclic Voltammetry, iR Effects and Applications," J. Electroanal. Chem., 161 (1984) 247.

H. Reller, I. Hallakoun-Feldstein and E. Gileadi. "Rotating Cylindrical Strip Microelectrodes," J. Electrochem. Soc., submitted.

Attila Szabo

D. Shoup and A. Szabo. "Hopscotch: An Algorithm for the Numerical Solution of Electrochemical Problems," J. Electroanal. Chem., 160 (1984) 1-17.

D. Shoup and A. Szabo. "Chronoamperometry at an Ensemble of Microdisk Electrodes," J. Electroanal. Chem., 160 (1984) 19-26.

D. Shoup and A. Szabo. "Influence of Insulation Geometry on the Current at Microdisk Electrodes," J. Electroanal. Chem., 160 (1984) 27-31.

A. Szabo, D. Cope, D. Tallman, P. Kovach and R.M. Wightman. "Chronoamperometric Current at Hemicylinder and Band Microelectrodes: Theory and Experiment," J. Electroanal. Chem., 217 (1987) 417-423.

A. Szabo. "Theory of the Current at Microelectrodes: Application to Ring Electrodes," J. Phys. Chem., in the press.

Dennis Tallman

D.J. Chesney, J.L. Anderson, D.E. Weisshaar and D.E. Tallman. "Evaluation of Kel-F-Graphite Electrodes as Detectors for Continuous Flow Systems," Analytica Chimica Acta, 124 (1981) 321.

D.E. Weisshaar, D.E. Tallman, J.L. Anderson. "Kel-F-Graphite Composite Electrode as an Electrochemical Detector for Liquid Chromatography and Application to Phenolic Compounds," Analytical Chemistry, 53 (1981) 1809.

D.E. Weisshaar and D.E. Tallman. "Chronoamperometric Response at Carbon-Based Composite Electrodes," Analytical Chemistry, 55 (1983) 1146.

D.E. Tallman and D.E. Weisshaar. "Carbon Composite Electrodes for Liquid Chromatography/ Electrochemistry: Optimizing Detector Performance by Tailoring the Electrode Composition," Journal of Liquid Chromatography, 6 (1983) 2157.

D.K. Cope and D.E. Tallman. "Calculation of Convective-Diffusion Current at a Strip Electrode in a Rectangular Flow-Channel: Inviscid Flow," Journal of Electroanalytical Chemistry, 188 (1985) 21.

D.K. Cope and D.E. Tallman. "Calculation of Convective-Diffusion Current at a Strip Electrode in a Rectangular Flow-Channel: Implications for Electrochemical Detection," Journal of Electroanalytical Chemistry, 205 (1986) 101.

S. Coen, D.K. Cope, and D.E. Tallman. "Diffusion Current at a Strip Electrode by an Integral Equation Method," Journal of Electroanalytical Chemistry, 215 (1986) 29.

A. Szabo, D.K. Cope, D.E. Tallman, P.M. Kovach, and R.M. Wightman. "Chronoamperometric Current at Hemicylinder and Band Microelectrodes: Theory and Experiment," Journal of Electroanalytical Chemistry, 217 (1987) 417.

D.K. Cope and D.E. Tallman. "Simplifications in the Theory of Step Experiments at Microelectrodes," Journal of Electroanalytical Chemistry, accepted for publication (1987).

S.L. Peterson and D.E. Tallman. "Kelsil: A New Precious Metal (Silver) Containing Composite Electrode for Analytical Voltammetry," Analytical Chemistry, to be submitted (1987).

R. Mark Wightman

R.M. Wightman. "Microvoltammetric Electrodes," Anal. Chem., 53 (1981) 1125A.

M.A. Dayton, A.G. Ewing and R.M Wightman. "Response of Microvoltammetric Electrodes to Homogeneous Catalytic and Slow Heterogeneous Charge-Transfer Reactions," Anal. Chem., 52 (1980) 2392-2396.

M.A. Dayton, J.C. Brown, K.J. Stutts and R.M Wightman. "Faradaic Electrochemistry at Microvoltammetric Electrodes," Anal. Chem., 52 (1980) 946-950.

A.G. Ewing, M.A. Dayton and R.M. Wightman. "Pulse Voltammetry with Microvoltammetric Electrodes," Anal. Chem., 53 (1981) 1842-1847.

M.A. Dayton, A.G. Ewing and R.M. Wightman. "Diffusion Processes Measured at Microvoltammetric Electrodes in Brain Tissue," J. Electroanal. Chem., 146 (1983) 189-200.

A.G. Ewing, J.C. Bigelow and R.M. Wightman. "Direct in vivo Monitoring of Dopamine Released from Two Striatal Compartments in the Rat," Science, 221 (1983) 169-171.

P.M. Kovach, A.G. Ewing, R.L. Wilson and R.M. Wightman. "In vitro Comparison of the Selectivity of Electrodes for in vivo Electrochemistry," J. Neuroscience Meth., 10 (1984) 215-227.

J. Millar, J.A. Stamford, Z.L. Kruk and R.M. Wightman. "Electrochemical, Pharmacological and Electrophysiological Evidence of Rapid Dopamine Release and Removal in the Rat Caudate Nucleus Following Electrical Stimulation of the Median Forebrain Bundle," Eur. J. Pharm., 109 (1985) 341-348.

W.G. Kuhr and R.M. Wightman. "Real-time Measurement of Dopamine Release in Rat Brain," Brain Res., 381 (1986) 168-171.

C. Amatore, R.S. Kelly, E.W. Kristensen, W.G. Kuhr and R.M. Wightman. "Effects of Restricted Diffusion at Ultramicroelectrodes in Brain Tissue. The Pool Model: Theory and Experiment for Chronoamperometry," J. Electroanal. Chem., 213 (1986) 31-42.

R.S. Kelly and R.M. Wightman. "Bevelled Carbon-Fiber Ultramicroelectrodes," Anal. Chim. Acta, 187 (1986) 79-87.

J.O. Howell and R.M. Wightman. "Ultrafast Voltammetry of Anthracene and 9,10-Diphenylanthracene," J. Phys. Chem., 88 (1984) 3915-3918.

J.O. Howell and R.M. Wightman. "Ultrafast Voltammetry and Voltammetry in Highly Resistive Solutions with Microvoltammetric Electrodes," Anal. Chem., 56 (1984) 524-529.

P.M. Kovach, W.L. Caudill, D.G. Peters and R.M. Wightman. "Faradaic Electrochemistry at Microcylinder, Band, and Tubular Band Electrodes," J. Electroanal. Chem., 185 (1985) 285-295.

K.W. Wehmeyer, M.R. Deakin and R.M. Wightman. "Electroanalytical Properties of Band Electrodes of Submicrometer Width," Anal. Chem., 57 (1985) 1913-1916.

C.A. Amatore, M.R. Deakin and R.M. Wightman. "Electrochemical Kinetics at Microelectrodes. Part I. Quasi-reversible Electron Transfer at Cylinders," J. Electroanal. Chem., 206 (1986) 23-36.

P.M. Kovach, M.R. Deakin and R.M. Wightman. "Electrochemistry at Partially Blocked Carbon-Fiber Microcylinder Electrodes," J. Phys. Chem., 90 (1986) 4612-4617.

M.R. Deakin, R.M. Wightman and C.A. Amatore. "Electrochemical Kinetics at Microelectrodes. Part II. Cyclic Voltammetry at Band Electrodes," J. Electroanal. Chem., 215 (1986) 49-61.

A. Szabo, D.K. Cope, D.E. Tallman, P.M. Kovach and R.M. Wightman. "Chronoamperometric Current at Hemicylinder and Band Microelectrodes: Theory and Experiment," J. Electroanal. Chem., 217 (1987) 417-423.

C.A. Amatore, B. Fosset, M.R. Deakin and R.M. Wightman. "Electrochemical Kinetics at Microelectrodes. Part III: Equivalency between Band and Hemicylinder Electrodes," J. Electroanal. Chem., submitted.

C.A. Amatore, M.R. Deakin and R.M. Wightman. "Electrochemical Kinetics at Microelectrodes. Part IV. Electrochemistry in Media of Low Ionic Strength," J. Electroanal. Chem., submitted.

K.R. Wehmeyer and R.M. Wightman. "Cyclic Voltammetry and Anodic Stripping Voltammetry with Mercury Ultramicroelectrodes," Anal. Chem., 57 (1985) 1989-1993.

K.R. Wehmeyer and R.M. Wightman. "Scan Rate Dependence of the Apparent Capacitance at Microvoltammetric Electrodes," J. Electroanal. Chem., 196 (1985) 417-421.

J.O. Howell, W.G. Kuhr, R.E. Ensman and R.M. Wightman. "Background Subtraction for Rapid Scan Voltammetry," J. Electroanal. Chem., 209 (1986) 77-90.

E.W. Kristensen, W.G. Kuhr and R.M. Wightman. "Temporal Characterization of Perfluorinated Ion-exchange Coated Microvoltammetric Electrodes for *in vivo* Use," Anal. Chem., submitted.

Index

A

AC equivalent circuit 59
AC methods
  Cole-Cole plots 55, 56, 203, 206
  impedance 52, 54, 137, 201-203, 207, 299
  voltammetry 83, 84, 115-118, 126, 150, 250
Acetonitrile 75, 117, 120, 121, 131, 134, 140, 141, 150, 153, 211, 213-216, 229, 232, 233, 269, 293, 296
Adiabatic charge transfer 308, 309
Adsorption 80, 127, 130, 153, 237, 246, 274, 288, 302
  monolayer 115, 128, 161
Advantages of
  ultramicroelectrodes 2, 3, 8, 9, 13, 85, 92, 100, 101, 117, 131, 143
Algorithms for simulations 58, 111
  hopscotch 143, 180
  hypogeometric representation 28
Alkane oxidation 14, 116, 140, 220, 293
Amperometry 2, 52, 111, 127, 136, 137, 143, 144, 146, 152, 160, 180, 188, 197, 199, 203, 242
Analysis 17
  approximate 67, 68, 110, 143, 182, 194, 196, 203
  conformal mapping 179, 180
  exact 11, 35-37, 43, 46, 47, 50, 54, 56, 141, 203, 265, 280
  finite difference 108, 109, 111, 143, 145, 177, 180, 188, 241
  hopscotch 143, 180
  simulations 111, 114, 119, 141, 143, 301-304, 307, 310-313
  solution for disk 22
  stochastic processes 14, 185, 202, 252
Anodic stripping analysis 137, 336, 338
Arrays 66, 70, 75, 76, 78, 94, 95, 100-104, 106, 107, 109-113, 117, 127, 133, 143, 145, 148, 149, 172, 173, 187-194, 240, 265, 275, 280, 338
Average surface concentration 24, 25, 27, 28, 35, 47, 51, 54, 60
Ascorbate 83, 178

B

Bessel function 19-21, 24, 25, 27, 41, 42, 43, 48, 50, 51, 60, 61
Boltzmann factor 305
Butler Volmer kinetics 28, 46, 55

C

Catalysis 90, 244, 299
Charge transfer reactions 55, 120, 123, 124, 171, 202, 221, 243, 244, 273, 289
Chromatography 110, 294
  capillary column 148
  detectors 8, 13, 107, 109, 145, 148, 173, 175, 188, 191, 210,

Index

295
HPLC 175, 335
Chronoamperometry 47, 48, 51, 137, 143, 146, 180, 197, 198, 203
Chronocoulometry 84, 117, 118, 337
Chronopotentiometry 2, 17, 40, 41, 46, 52, 136
  transition time 3, 52, 197, 198
Clusters 252
Cole-Cole plots 55, 56, 203, 206
Complementary function 41, 50
Conductivity 148, 171, 182, 184, 284, 286, 294, 295
Constant flux assumptions 141
Constant concentration assumptions 23, 24, 141
Convection layer 169
Coordinates for diffusion models 152, 160, 304
  circular 17, 18, 81, 117, 149, 207, 247, 265
  spherical 152
Copolymers 160, 162
Coupled chemical reactions 2, 3, 114, 120, 121, 185, 274
  CE 48, 85
  EC 38, 39, 85, 170
  ECE 121
Current
  capacitative (charging) 2, 4, 139, 154, 188, 243, 293, 294, 302
  constant flux approximation 141
  constant concentration approximation 141
  distribution 171, 186, 194, 246, 286, 306, 309, 311, 312
  faradaic 86, 125, 186, 245, 276, 281, 294, 334
  functions 3, 11, 12, 73, 100, 118, 129, 146, 148, 154, 190, 214, 217, 218, 235, 295, 304, 306
  limiting 29, 35, 79, 80, 121, 133, 172, 198, 200, 203, 213, 214, 218, 221, 222, 268, 270, 291, 294, 297
Current (cont.)
  migration 3, 9, 134, 148, 172, 186, 284, 286, 288, 289, 307, 308, 336
  nonfaradaic 153
  nonuniform 96, 121, 170, 171, 186, 246, 286
  tertiary 11
  transient 1, 2, 35, 40, 46, 84, 117-119, 126, 139, 150, 172, 238, 249, 251, 308
Current amplifiers 118, 225, 226, 228, 231, 235
Cyclic voltammetry 58, 115, 139, 177, 213, 333, 335
  rapid scan 154

D

Debye screening 246, 286, 312, 314
Detection 109-111, 118, 120, 126, 145, 175, 178, 185, 188, 189, 191, 210, 235, 241, 295, 335
  intercellular 14
  metal ion 158, 160
Dielectric solvents, measurements in 273, 298, 299
Differential equations 18, 19, 28, 40, 48, 108, 146, 180
Differential pulse polarograph 84, 118, 150, 152, 161, 162, 165, 198
Diffuse layer 172

Diffusion
  cylindrical 67, 83, 95,
    104, 106, 119, 127,
    147, 177, 181, 196,
    265, 292, 337, 339
  lateral 111, 128, 170
  nonplanar 169, 170, 177,
    289, 292, 334, 339
  overlap 143, 172, 173
  planar 6, 48, 95, 143,
    145-147, 170, 174,
    183, 201
  radial 34, 139, 265, 268
  spherical 8, 15, 70, 85,
    116, 136, 143, 147,
    152, 174, 175, 181,
    183, 203, 242, 243
  steady state 1, 3, 11, 18,
    22, 38, 40, 46, 60, 84,
    101, 175, 181, 188, 191,
    197, 199, 202, 222, 242,
    248, 281, 294
Diffusion layer 109, 110,
  143, 169, 170, 172, 173,
  177, 181, 189, 190
Dimensionless variables and
  parameters 26, 30, 31, 34,
  43, 45, 46, 55, 56, 108
Dimensions 3, 4, 12, 14, 118,
  186, 187, 189-191, 207,
  211, 223, 247, 263, 265,
  280, 333, 338
  characteristic 67, 69-71,
    76, 84, 91, 96-98,
    101, 102, 105, 118,
    124, 152, 169, 171,
    248, 281, 283, 284,
    289
Disadvantages of
  ultramicroelectrodes 109,
  276, 277
Discontinuous integrals 18,
  21, 47, 60, 61
Dispersions 17, 140
  bipolar 5, 85, 242, 245,
    299
  particulate suspensions
    274

Double layer considerations
  55, 73, 202, 246, 286,
  293, 294, 301, 302, 312

E

Electrocatalysis 127, 299
Electron transfer reactions
  32, 80, 110, 111, 119-121,
  123, 130, 135, 141, 147,
  150, 177, 189, 191, 221
  246, 270, 273, 274, 293,
  295, 301-303, 306, 309,
  310, 312, 313
Electron microscopy 97, 98,
  220, 265-267
Epoxy resin 67-71, 74-78,
  84, 92-94, 96-98, 102,
  104-106, 117, 118, 136,
  220, 221, 290, 294, 337
Eutectic mixtures 120, 293

F

Finite difference simulations
  108, 109, 111, 143, 145,
  177, 180, 188, 241
Flow cells 107, 109-113, 125,
  145, 150, 187-194, 210,
  235, 243, 297, 335
Fractal dimensions 66
Frequency factor 305-307, 309

G

Galvanostatics 42, 52
Gas phase electrochemistry
  140, 184, 242-244,
  293-295, 303

H

Homogeneous reactions 2, 3,
  114, 120, 121, 185, 274

Hydrogen reactions 4-6, 10,
    11, 83, 121, 215, 237, 248,
    253, 307
Hypogeometric function 25-28

I

Impedance 137, 201-203, 207,
    299
Impurity 120
Infrared spectroscopy 211,
    215-217
Initial state 305-307, 311
Instrumentation for
    ultramicroelectrode
    measurements 1, 105, 117,
        119, 210, 263
    current follower 226, 235
    faraday cage 295
    picoammeter 118, 225
    potentiostat 115, 118,
        225, 226, 228, 275,
        339
Insulators 5, 66-68, 70, 72,
    80, 81, 84, 96, 105, 171,
    201, 243, 276, 290, 292,
    294, 295
Ion conduction 295
Ion exchange reactions 6,
    158, 161

K

Kinetics 11, 18, 23, 30, 124,
    126, 128, 132, 182, 198,
    199, 201, 242, 247, 252,
    284, 286, 288
    electron transfer 3-5, 8,
        9, 80, 110, 111, 119-
        121, 123, 130, 135,
        141, 147, 150, 177,
        189, 191, 221, 246,
        270, 273, 274, 293,
        295, 301-303, 306,
        309, 310, 312, 313
    homogeneous 2, 3, 114,
        120, 121, 185, 274
    rate constants 3, 132,
        141, 172, 232, 303
Laplace space 40, 41, 47, 48,
    146
Linear sweep voltammetry 52,
    136, 197, 199
Lord Kelvin 23

M

Mass transport 1-4, 7-9, 11,
    12, 15, 30, 35, 86, 130,
    139, 141, 143, 169-171,
    186, 196, 273, 275, 276,
    293, 334, 335
    constant flux 141
    constant concentration
        43, 58, 141
Mass transfer coefficient 15,
    23, 27, 28, 35
    constant flux 18, 22-25,
        28-32, 35-37 48, 141
    constant concentration
        18, 23, 29-32, 58,
        141
Mercury 15, 91, 115, 125,
    136, 137, 154, 248, 249,
    276, 336-338
    dropping mercury electrode
        91
Metropolis method 305
Migration 3, 9, 134, 148,
    172, 186, 284, 286, 288,
    289, 307, 308, 336
Modified electrodes 8, 101,
    159, 245
Monolayer adsorption 115,
    128, 161

N

Nonaqueous electrochemistry
    75, 211, 293
Nucleopore membranes 69, 95

O

Operational amplifier 226, 228, 235
Overpotential 29, 30

P

Particulate suspensions 274
Photolithography 13, 100, 102, 103
Poisson processes 185, 186, 312
Polarization curves or plots 28, 29
Polarography 2, 82, 91, 115, 268, 270, 295
   differential pulse 84, 116, 150, 152, 161, 162, 165, 198
Pre-exponential factor 302

R

Reference electrodes 118, 134, 284, 294, 337
Resistance 1, 2, 8, 12, 58, 74, 75, 115-117, 119, 120, 123, 134, 135, 139, 150, 171, 172, 173, 186, 202, 203, 213, 226, 232, 237, 243, 270, 276, 284, 285, 293, 294-296, 336
Reversible 116, 117, 119, 141, 147, 152, 160, 161, 191, 298

S

Saddle point 303, 304, 306-308, 310, 311
Sensors 8, 100, 131, 148, 182, 185, 240, 275, 298
   brain tissue 175, 177, 178
   chromatographic 107, 109, 120, 140, 295
   in vivo 65, 67, 81, 153, 173, 175, 184, 333
Simulations of analysis 111, 114, 119, 141, 143, 301-304, 307, 310-313
Single electron counting 134, 185, 301, 303
Solutions 4, 11, 15, 115, 119-121, 124, 145, 146, 152, 154, 313, 335
   gas phase 2, 9, 10, 140, 242-244, 293-295, 303
   homogeneous 2, 3, 114, 120, 121, 274
   low temperature 8, 242, 293
   resistance 1, 2, 8, 12, 74, 75, 115-117, 119, 120, 123, 134, 135, 139, 150, 171, 172, 173, 186, 202, 203, 213, 226, 232, 237, 243, 270, 276, 284, 285, 293, 294-296, 336
   solid, 2, 7, 8, 13, 89, 100, 110-112, 131-135, 244, 293, 294
Steam 10, 11
Stochastic processes 14, 185, 202, 252
Supporting electrolyte 2, 8, 117, 120, 125, 131, 140, 150, 210, 211, 213-215, 217, 240, 286, 288, 293, 299, 336
Symmetry 8, 169, 173, 175, 181, 182
Synthetic applications 6, 228

T

Transcendental functions 20
Transition time 3, 45, 52, 197, 198

Transition state 303, 305-311
Tunnelling 66, 70, 244, 306-310

U

Ultramicroelectrode
  advantages 2, 3, 8, 9, 13, 85, 92, 100, 101, 117, 131, 143
  band 7, 66, 67, 70, 76, 78, 98, 99, 141, 123, 124, 146, 150-153
  cylinder 66-70, 76, 91-94, 97, 98, 126, 146, 153, 265, 273, 276
  dimensionality 187, 274, 280-282, 284, 285
  dimensions 3, 4, 12, 14, 33, 34, 40, 67, 69-71, 76, 84, 91, 96-98, 101, 102, 105, 118, 124, 152, 169, 171, 186, 187, 189-191, 207, 211, 223, 247, 248, 263, 265, 280, 281, 283, 284, 289, 333, 338
  disadvantages 109, 276, 277
  disk 11, 12, 14, 17, 20, 21, 24, 26-28, 33, 34, 39, 48, 51, 58, 61, 66-70, 73, 76, 80, 81, 91-96, 102, 117-119, 121, 126, 127, 129, 132, 136, 137, 141, 143, 149, 150, 154, 242, 243, 249, 263-265, 273, 276, 289, 291-294, 298, 338
  dispersion 6, 5, 85, 140, 242, 245, 299,
  ensembles 117, 149, 150, 247, 302-306, 339
  fiber 35, 66, 68, 73, 74, 76-78, 80-83, 93-98, 137, 152, 294, 337, 338
  inlaid 118, 276, 289, 291
  mercury 15, 91
  particulate suspensions 274
  ring 18, 26, 30, 33-35, 37, 58, 68-70, 84, 96-98, 117, 129, 141, 144, 243, 246, 273, 276, 280, 294
  semiconductor 5, 7, 68, 96
  spheres 8, 15, 38, 40, 66, 68, 70, 85, 91, 116, 121, 136, 143, 147, 152, 242, 243, 273-276
  zeolite 70, 71, 85, 299
Uncompensated resistance 115, 270

V

Vacuum applications 73, 76, 92, 94, 104, 106, 192, 244, 245
Voltammetry 52, 83, 84, 115-118, 120, 125, 126, 134-137, 139, 148-150, 153, 154, 158, 161, 177, 197-199, 211, 213, 220-223, 228, 230, 231, 288, 290, 291, 333, 335, 336, 338

W

Warburg impedance 201-203

Z

Zeolite 70, 71, 85, 299